4TH EDITION
INTRODUCTORY STATISTICS

MATH 10041 · ACTIVITY BOOK

DEPARTMENT OF MATHEMATICAL SCIENCES
KENT STATE UNIVERSITY

VAN-GRINER
LEARNING

Introductory Statistics

Math 10041 Activity Book
Department of Mathematical Sciences
Kent State University
4th Edition

Copyright © by Department of Mathematical Sciences

Photos and other illustrations are owned by Van-Griner Learning or used under license.
All products used herein are for identification purposes only, and may be trademarks or registered trademarks of their respective owners.

All rights reserved. No part of this book may be reproduced or transmitted in any form or by any means, electronic or mechanical, including photocopying, recording or by any information storage and retrieval system, without written permission from the author and publisher.

Printed in the United States of America
10 9 8 7 6 5 4 3 2 1
ISBN: 978-1-64565-393-6

Van-Griner Learning
Cincinnati, Ohio
www.van-griner.com

President: Dreis Van Landuyt
Senior Project Manager: Maria Walterbusch
Customer Care Lead: Lauren Wendel

Reed 65-393-6 Su23
332434
Copyright © 2024

MATH 10041 Introductory Statistics ACTIVITY BOOK

4th edition

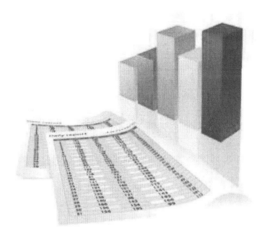

Department of Mathematical Sciences

Kent State University

Table of Contents

Chapter 1: Introduction to Data 1

Chapter 2: Picturing Variation with Graphs 13

Chapter 3: Numerical summaries of Center and Variation 29

Chapter 4: Regression Analysis: Exploring Associations between Variables 53

Chapter 5: Modeling Variation with Probability 73

Chapter 6: Modeling Random Events: The Normal and Binomial Models 103

Chapter 7: Survey Sampling and Inference 119

Chapter 8: Hypothesis Testing for Population Proportions 143

Chapter 9: Inferring Population Means 161

Chapter 1

Introduction to Data

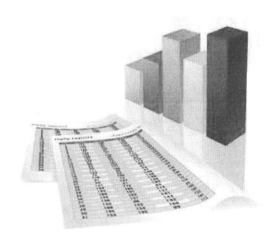

Developing a Class Survey

We are going to learn about students in this class by gathering some data on a survey. You get to contribute to this survey, so this is your opportunity to think about questions that will produce some interesting data about your classmates!

In your small group, discuss and write five, carefully worded items for a class survey. We want these items to be clear and unambiguous, so try them out on yourselves before writing down the final version, and see if modifications are needed.

But, before you begin, there are some rules:

1. Make sure at least one of your five questions asks for **numerical** or **quantitative** data (e.g., a number).

2. Make sure at least one of your five questions asks for binary data (e.g., yes/no, M/F, etc).

3. Make sure at least one of your five questions asks for **categorical** or **qualitative** data (e.g., names, categories, etc.).

4. Don't produce boring questions! Try to create five interesting items that we can use to find out what students in this class are like.

5. Write or type your questions on the next page and submit them before you leave class.

(OVER)

Reference

Garfield, J., Zieffler, A., & Lane-Getaz, S. (2005). EPSY 3264 Course Packet, University of Minnesota, Minneapolis, MN.

Table Number:_____   Group Name: _____

Group Members:_____   _____   _____

Developing a Class Survey

Five questions:

In your small group, discuss and write five, carefully worded items for a class survey. We want these items to be clear and unambiguous, so try them out on yourselves before writing down the final version, and see if modifications are needed. Please TYPE or WRITE THEM below.

Numerical (Quantitative):

Binary:

Categorical (Qualitative):

Table Number:_____   Group Name: _____

Group Members:_____   _____   _____

Gettysburg Address

One of the most important ideas in statistics is that we can learn a lot about a large group (called a population) by studying a small piece of it (called a sample). Consider the population of 268 words in the following passage:

> *Four score and seven years ago, our fathers brought forth upon this continent a new nation: conceived in liberty, and dedicated to the proposition that all men are created equal.*
>
> *Now we are engaged in a great civil war, testing whether that nation, or any nation so conceived and so dedicated, can long endure. We are met on a great battlefield of that war.*
>
> *We have come to dedicate a portion of that field as a final resting place for those who here gave their lives that that nation might live. It is altogether fitting and proper that we should do this.*
>
> *But, in a larger sense, we cannot dedicate, we cannot consecrate, we cannot hallow this ground. The brave men, living and dead, who struggled here have consecrated it, far above our poor power to add or detract. The world will little note, nor long remember, what we say here, but it can never forget what they did here.*
>
> *It is for us the living, rather, to be dedicated here to the unfinished work which they who fought here have thus far so nobly advanced. It is rather for us to be here dedicated to the great task remaining before us, that from these honored dead we take increased devotion to that cause for which they gave the last full measure of devotion, that we here highly resolve that these dead shall not have died in vain, that this nation, under God, shall have a new birth of freedom, and that government of the people, by the people, for the people, shall not perish from the earth.*

(a) Select a sample of ten *representative* words from this population by circling them in the passage above.

The authorship of several literary works is often a topic for debate. Were some of the works attributed to William Shakespeare actually written by Francis Bacon or Christopher Marlowe? Which of the anonymously published Federalist Papers were written by Alexander Hamilton, which by James Madison, which by John Jay? Who were the authors of the writings contained in the Bible? The field of "literary computing" began to find ways of numerically analyzing authors' works, looking at variables such as sentence length and rates of occurrence of specific words.

The above passage is, of course, Lincoln's Gettysburg Address, given November 19, 1863 on the battlefield near Gettysburg, PA. In characterizing this passage, we could have asked you to examine every word. Instead, we asked you to look at a sample of the words of the passage. We are considering this passage a **population** of words, and the 10 words you selected are considered a **sample** from this population. In most studies, we do not have access to the entire population and can only consider results for a sample from that population. The goal is to learn something about a very large population (e.g., all American adults, all American registered voters) by studying a sample. The key is in carefully selecting the sample so that the results in the sample are **representative** of the larger population (i.e., has the same characteristics).

> The **population** is the entire collection of observational units that we are interested in examining. A **sample** is a subset of observational units from the population. Keep in mind that these are objects or people, and then we need to determine what variable we want to measure about these entities.

(b) Do you think the ten words in your sample are representative of the 268 words in the population? Explain briefly.

(c) Record the length for each of the ten words in your sample:

Word	1	2	3	4	5	6	7	8	9	10
# letters										

(d) Determine the average (mean) number of letters in your sample (the ten words).

(e) List the average (mean) number of letters of the samples of each of the other two groups at your table and also at one neighboring table. List those averages (means) here:

_____ _____ _____ _____ _____ _____

(f) The population average number of letters for all 268 words is 4.295 letters. Were most of the samples' means *near* the population mean? Explain.

(g) For how many groups did the sample average exceed the population average? What proportion of the six groups is this?

(h) Explain why this sampling method (asking people to choose ten "representative" words) is **biased** and how this bias is exhibited. Also identify the *direction* of the bias. In other words, does the sampling method tend to overestimate or underestimate the average length of the words in the passage?

A **simple random sample** (SRS) gives every observational unit in the population the same chance of being selected. In fact, it gives every sample of size *n* the same chance of being selected. In this example we want every set of ten words to be equally likely to be the sample selected. While the principle of simple random sampling is probably clear, it is by no means simple to implement.

(i) Use the app in the link below to generate 6 random samples of size 10, one at a time, and write the mean of each sample here:

_____ _____ _____ _____ _____ _____

You'll need to click on "Show Sampling Options" ☑ in order to choose the number of samples and the sample size. Choose one at a time and note the mean of each sample.

Show Sampling Options: ☑

Number of Samples: 1

http://www.rossmanchance.com/applets/OneSample.html?population=gettysburg

(j) Compare the results of part (i) to part (e). Comment on the randomness and reliability of your samples in part (e) to those in part (i).

(k) Compare the means you obtained from the 6 samples from part (e) and part (i) to the population mean. Justify and explain the differences you observe from part (e) and part (i).

Other Sampling Methods

(l) A **systematic sample** is obtained by selecting every k^{th} individual from the population. The first individual selected corresponds to a random number between 1 and k.

    i.   Choose an arbitrary value for k and write it here:_____

    ii.   Use the random number generator on your calculator to choose a value between 1 and k. Suppose this value is 4. You will then start with the 4th word in the Gettysburg address and choose every k^{th} word for your sample. Write your value of k here:_____.

    iii.   Underline each word in your systematic sample.

    iv.   Find the mean of your sample and write it here: _____

    v.   How does this mean compare to the population mean?

(m) A **stratified sample** is obtained by separating the population into non-overlapping groups called *strata* and then obtaining a simple random sample from each stratum. The individuals within each stratum should be similar in some way. How might you collect a stratified sample of ten words from the Gettysburg address? Write your plan here:

(n) A **cluster sample** is obtained by selecting all individuals within a randomly selected collection or group of individuals. How might you collect a cluster sample of ten words from the Gettysburg address? Write your plan here:

Table Number:_____   Group Name: _____

Group Members:_____   _____   _____

Random Sampling Methods

Key Understanding: Every member is equally likely to be chosen

---

Simple Random Sampling – Use a random number generator or a random number table to determine which members are chosen.

Cluster Sampling – Population is divided into clusters reflecting the variability within population and then a certain number of clusters are randomly selected and every member of the selected cluster is surveyed.

Systematic Sampling – Sampling every k^{th} member of a population after randomly determining the first individual by selecting a random number between 1 and k.

Stratified Random Sampling – The population is divided into two or more groups (strata) according to some criterion and subsamples are randomly chosen from each strata.

Identify the type of sampling used in each of the following vignettes and explain your choice:

1. Ohio wants to gauge public reaction to a proposed tax increase. They randomly select individuals from rural areas, randomly select individuals from suburban areas, and randomly select individuals from urban areas.
   Type of sampling:_____
   Explain:

2. Bauman's Orchard wants to determine the approximate yield of its peach trees. They divide the orchard into 25 subsections and randomly select 4. They sample all the trees within the 4 subsections to approximate the yield of the orchard.   Type of sampling: _____
   Explain:

3. Every hundredth hamburger manufactured is checked to determine its fat content.
   Type of sampling: _____
   Explain:

4. Design consultants are selected using random numbers in order to determine annual salaries.
   Type of sampling: _____
   Explain:

5. A survey regarding download time on a certain web site is administered on the internet by a market research firm to anyone who is willing to respond to the survey. Type of sampling: _____
   Explain:

Table Number:_____   Group Name: _____

Group Members:_____   _____   _____

Beethoven and Intelligence

Part I: Designing an observational study

1. With your group, design an observational study to compare the intelligence of infants who listen to Beethoven and those who don't. Consider an infant as being age birth to one year. Explain in significant detail how you would run this study. Refer to pages 14 – 19 in your text as needed.

> STATUS QUO: *There is no difference in intelligence of infants who listen to Beethoven and those that do not.*
>
> WHAT WE THINK IS TRUE: *There is a difference in intelligence between infants who listen to Beethoven and those who do not.*

2. Assuming that infants who listen to Beethoven do have higher intelligence scores, would you be able to conclude that listening to Beethoven affects intelligence? Explain

3. Is there a confounding variable that might affect the outcome of the study?

Part II: Designing a controlled experiment

4. With your group, design a controlled experiment to compare the intelligence of infants who listen to Beethoven and those who don't. Consider an infant as being age birth to one year. Explain in significant detail how you would run this study. Refer to the gold standard for experiments on p. 19 in your text.

5. Each group is to critique the study of another group. Share your file with another group at your table. You might want to just shift one group over at the table. Please type your critique in red or blue and list your table and group number by your critique. Comment on each of the elements of a good research design. Which were met in the study? Which weren't? Explain. Return the study to its original writers.

6. Comment on the critique from the other group. Do you agree or disagree with their comments? Would your study be a better one if you followed their suggestions? Explain.

Chapter 2

Picturing Variation with Graphs

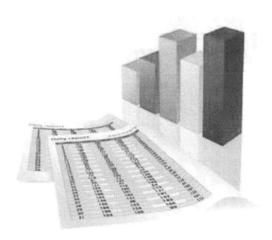

Table Number:_____ Group Name: _____

Group Members:_____ _____ _____

Stem and Leaf Plots

A random group of 22 college students took a standardized aptitude test. Their scores were:

262	243	228	231	236
258	247	249	262	237
259	250	265	253	232
249	258	235	269	244
263	255			

Construct a stem-and-leaf display for this data. Be sure to include a key. Summarize the important features of the numerical distribution.

Table Number:_____   Group Name: _____

Group Members:_____   _____   _____

Constructing a Two-Way Table

Two rival television stations conducted a random survey of residents in the area. The data table below summarizes the findings.

Support WNEO (Ch 49)	
Yes	887
No	613

Support WVIZ (Ch 25)	
Yes	623
No	877

Support both
365

Support neither
355

1. Construct a two-way table below to organize the data.

2. What percent of the participants in the survey support WNEO only? Please show your set up as well as your answer as a percent.

3. What percent of the participants in the survey support WVIZ only? Please show your set up as well as your answer as a percent.

Table Number:_____   Group Name: _____

Group Members:_____   _____   _____

Distinguishing Distributions

Refer to the dot plots on the document, DOT PLOTS FOR "DISTINGUISHING DISTRIBUTIONS" ACTIVITY on pp. 21-22 in your activity book or in the *GW6 Distinguishing Distributions* assignment in our Canvas course to answer the questions below. Each dot plot depicts the distribution of hypothetical exam scores in various classes.

1. For classes A, B and C, what is the main characteristic that distinguishes these three graphs from each other? What might explain this difference?

2. What is the main characteristic that distinguishes the distributions of exam scores in classes D, E, and F? What might explain these differences?

3. What is the main characteristic that distinguishes the distributions of exam scores in classes G, H, and I? What might be an explanation for the distinguishing feature you find?

4. What strikes you as the most distinctive features of the distribution of exam scores in class J? What might be an explanation for this characteristic?

5. What strikes you as the most distinctive feature of the distribution of exam scores in class K? What might be an explanation for this characteristic?

6. Look again at the graph for class D. If you wanted to tell someone about how the students did on this exam, what would you say?

7. What if we tried to look at the "bulk" of the data? Now, what would you say?

8. What about the graph for class E?

9. What if you were told that most exam scores for a class were between 65 and 85, and that the overall range of scores for that class was between 30 and 100? Can you imagine what that distribution would look like? Sketch it below.

Reference

Rossman, A., & Chance, B. (2002). A data-oriented, active-learning, post-calculus introduction to statistical concepts, applications, and theory. In B. Phillips (Ed.), *Proceedings of the Sixth International Conference on Teaching of Statistics,* Cape Town. Voorburg, The Netherlands: International Statistical Institute. Retrieved September 28, 2007, from http://www.stat.auckland.ac.nz/~iase/publications/1/3i2_ross.pdf

Dot Plots for "Distinguishing Distributions" Activity

DOT PLOTS FOR QUESTION #1:

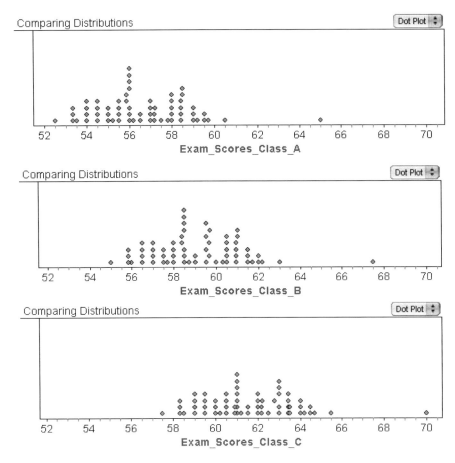

DOT PLOTS FOR QUESTION #2:

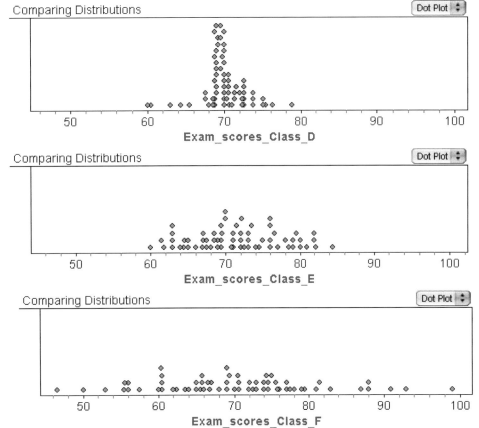

DOT PLOTS FOR QUESTION #3:

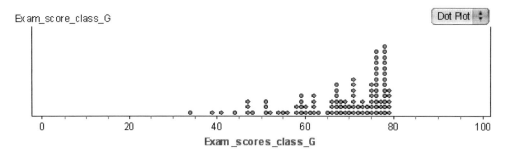

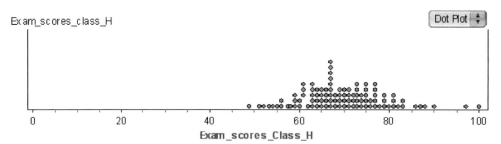

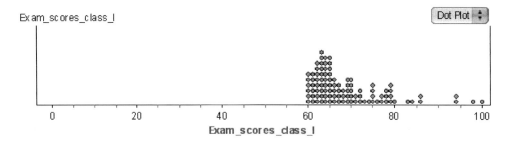

DOT PLOT FOR QUESTION #4:

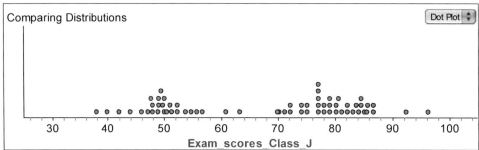

DOT PLOT FOR QUESTION #5:

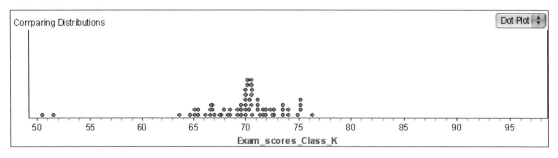

Table Number:_____   Group Name: _____

Group Members:_____   _____   _____

Age of Presidents at Inauguration

How old are presidents at their inauguration? Was Bill Clinton, at age 46, unusually young?

Enter the following ages of presidents at inauguration into StatCrunch. Label the column **Age.** This data is also available on our Blackboard course as an Excel file so that you can upload the data to Stat Crunch directly.

President	Age	President	Age	President	Age
Washington	57	Buchanan	65	Coolidge	51
J. Adams	61	Lincoln	52	Hoover	54
Jefferson	57	A. Johnson	56	F.D.Roosevelt	51
Madison	57	Grant	46	Truman	60
Monroe	58	Hayes	54	Eisenhower	61
J.Q. Adams	57	Garfield	49	Kennedy	43
Jackson	61	Arthur	51	L. Johnson	55
Van Buren	54	Cleveland	47	Nixon	56
W.H.Harrison	68	B. Harrison	55	Ford	61
Tyler	51	Cleveland	55	Carter	52
Polk	49	McKinley	54	Reagan	69
Taylor	64	T.Roosevelt	42	G. Bush	64
Fillmore	50	Taft	51	Clinton	46
Pierce	48	Wilson	56	G.W. Bush	54
		Harding	55	B. Obama	47

1. Construct a histogram on StatCrunch , choosing **Frequency** under the **Type** drop-down menu. Let StatCrunch decide upon the number of bins. Leave that field blank. Be sure to name the histogram and to label the axes. Copy and paste your histogram here:

2. a) Is the distribution symmetric or skewed? Explain

    b) Is the data concentrated in one or two bins or is it evenly spread? If it is concentrated, what bins are the areas of concentration? Does this seem to characterize a "typical" value? Explain your reasoning.

3. Write a brief description of the distribution you created, describing its shape, center, and spread.

4. Now create another StatCrunch histogram, but this time let the bin width = 3. Copy and paste it in the space below, then describe how this new histogram differs from your first one. Describe its shape, center, and spread.

5. Experiment with two other bin widths (but not a bin width of 1), copying and pasting your histograms below. How are your new histograms different from the original? How might the differences affect your ability to determine the average age of the presidents?

Table Number:_____   Group Name: _____

Group Members:_____   _____   _____

Matching Histograms to Variable Descriptions

Part I: Warm-Up Example

Below are two histograms (the scales have been left off intentionally). One corresponds to the age of people applying for marriage licenses, the other corresponds to the last digit of a sample of social security numbers. Label the horizontal axis of each graph with either AGE or LAST DIGIT, and write an explanation of why you made each choice.

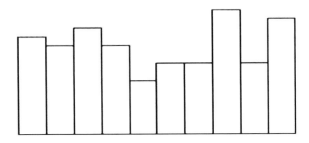

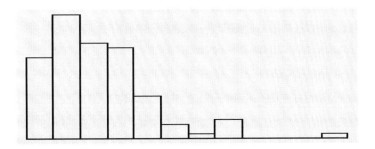

(OVER)

Part II

Match each of the following variable descriptions with one of the histograms given below. *Give your reason for each of your matches.* Discuss your answers and justifications with other students in your group.

1. The height of college students in a class. Write the letter under the graph here: _____
   Explain:

2. Number of siblings of college students in a class. Write the letter under the graph here: _____
   Explain:

3. Amount paid for last haircut by college students in a class. Write the letter under the graph here: _____
   Explain:

4. Students' guesses of their professor's age on the first day of class. Write the letter under the graph here: _____
   Explain:

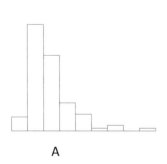

A

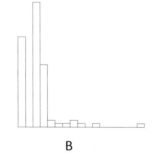

B

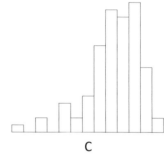

C

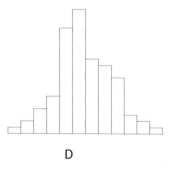

D

References

Rossman, A., Chance, B., & Lock, R. (2001). *Workshop statistics: Discovery with data and Fathom*. Emeryville, CA: Key College Publishing.
Scheaffer, R.L., Watkins, A. Witmer, J., & Gnanadesikan, M., (2004a). *Activity-based statistics: Instructor resources* (2nd edition, Revised by Tim Erickson). Key College Publishing

Table Number:_____   Group Name: _____

Group Members:_____   _____   _____

Oil Consumption

The **total world-wide** barrels of oil used per day is **82,234,918**. The **top five** countries in terms of oil consumption per day are listed in the table below. Complete the table by finding the number of barrels per day consumed by the "Other" countries.

Country	Barrels per day
United States	20,730,000
China	6,543,000
Japan	5,578,000
Germany	2,650,000
Russia	2,500,000
Other	?

1. Using StatCrunch, create two different graphical displays for world-wide oil consumption. Remember to include a category entitled "other" in your graphs so that you include the **total** world-wide production, not just the five countries listed in the table. Copy your graphs and paste them below.

2. Why might we use one type of graph instead of the other? What are some key differences that distinguish the different types of graphs? Explain how each gives different information or a different impression.

3. What percent of worldwide oil does the U.S consume? _____ Does China consume? _____

Chapter 3

Numerical Summaries of Center and Variation

Chapter 3 31

Table Number:_____   Group Name: _____

Group Members:_____   _____   _____

What is Typical?

Part I: Making Predictions

For each of the following variables measured on the *Student Survey* (your section), make a prediction for a ***typical value*** *for all students enrolled in your statistics class this term*. A typical value is a single number that summarizes the class data for each variable.

1. Write that prediction in the *First Prediction* column.

Attribute from Student Survey	First Prediction	Revised Prediction from DotPlot	Statistics from *StatCrunch*	
			Mean	Median
Age				
College credits registered for this semester				
Number of states visited				
Height (in inches)				
Number of siblings				
Hours a week spent studying				
Hours a week spent working at a paying job				

Open MyMathLab and click on the *StatCrunch* link in the menu on the left. Click on the link "Stat Crunch Website,"

View the data sets from your textbook in StatCrunch.

Visit the StatCrunch website to perform complex analyses with the StatCrunch statistical software, search shared data sets, take online surveys, and more.

then "**My Groups.**" Click on our group, then under the "Preview Data," click on "Class data."

2. Now use *StatCrunch* to create dot plots of each variable to see if your original predictions seem reasonable. Based on the dotplots, make revised predictions for the typical value for each of the variables.

3. To make a dot plot in *StatCrunch*, click on the **Graph** button at the top of the spreadsheet, choose Dotplot.

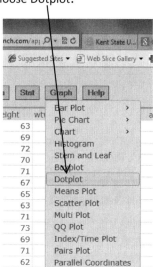
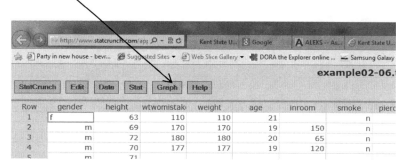

4. Choose the column of interest, label the axes, then click on "Compute!" at the bottom of the page.

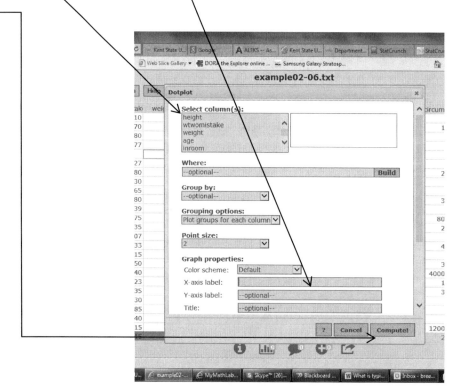

5. Write these new predictions in the *Revised Prediction* column in the table.

Part II: Test Your Conjectures

Use *StatCrunch* to find the mean and median for each of these variables. (Follow the directions below.)

1. Click the Stat button.

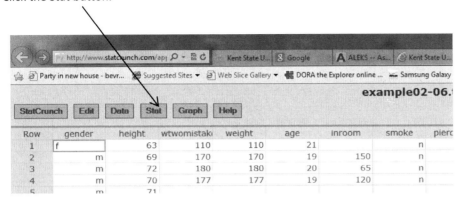

2. Click on "Summary Stats," then "Columns." Choose the columns of interest. You can select more than one by holding down the Ctrl key on your keyboard.

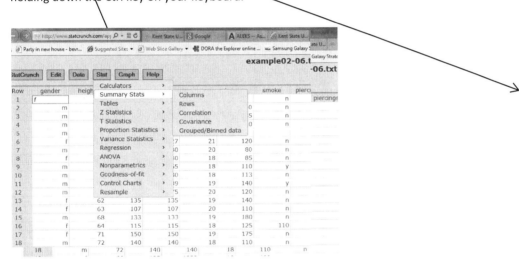

3. Click on the stats you want. You can choose more than one by holding down the Ctrl key on your keyboard
4. Click "Compute."

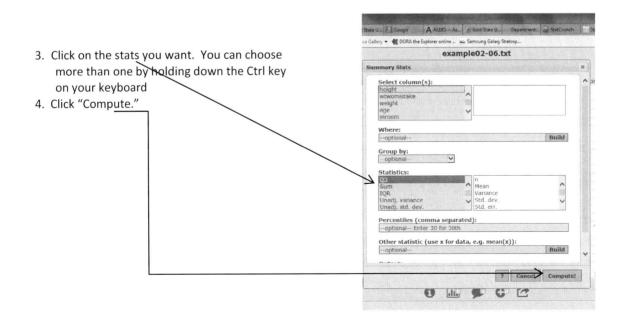

6. How close were your *revised predictions* to the "typical" values produced in *StatCrunch*? For which attributes were your predictions most accurate?

7. What was most surprising to you? Why?

8. In general, were your *revised predictions* closer to the means or medians?

Things to Consider

- How close were your predicted typical values?

- Which measure of center were your guesses closer to, the mean or median?

- What information do means and medians tell us about a distribution?

- How do we decide whether to use the mean or median to summarize a data set?

- In statistics, what do we mean by what is typical?

Reference

Adapted from AIMS http://www.tc.umn.edu/~aims/aimstopics.htm

Table Number:_____   Group Name: _____

Group Members:_____   _____   _____

Standard deviation – The Basics

1. Find the Deviations. Given $\bar{x} = 2.8$

	Deviations from $\bar{x}$	Deviations SQUARED
0		
5		
6		
0		
2		
4		

2. The top speeds (in mph) for a sample of five new automobile brands are listed below. Calculate the standard deviation of the speeds.

	Deviations from $\bar{x}$	Deviations SQUARED
160		
125		
190		
185		
105		

3. The book cost (in dollars) for one semester's books are given below for a sample of five college students. Calculate the sample standard deviation of the book costs.

	Deviations from $\bar{x}$	Deviations SQUARED
200		
130		
400		
500		
345		

4. Each of the following distributions have a mean of 60. Without calculating the standard deviation, decide which of the following data sets has the largest standard deviation. Which has the smallest? Explain to your group members WHY you think so. Then check with your graphing calculator.

   a)   60 60 60 60 60 60
   b)   10 20 25 95 100 110
   c)   50 50 55 65 70 70

The following list shows the age at appointment of U.S. Supreme Court Chief Justices appointed since 1900. Use the data to answer the question.

Last Name	Age
White	65
Taft	63
Hughes	67
Stone	68
Vinson	56

Last Name	Age
Warren	62
Burger	61
Rehnquist	61
Roberts	50

5. The US Supreme Court Chief Justice data was used to create the following output in an Excel spreadsheet. Choose the statement that best summarizes the variability of the dataset.

Column1	
Standard Error	1.864217971
Standard Deviation	5.592653912
Sample Variance	31.27777778
Kurtosis	1.177934931
Skewness	-1.064306225
Range	18
Minimum	50
Maximum	68
Sum	553
Count	9

   a. The ages of most of the US Supreme Court Chief Justices are between 50 and 68 years.
   b. The age of most of the US Supreme Court Chief Justices since 1900 are within 5.6 of the mean age.
   c. The age of most of the US Supreme Court Chief Justices since 1900 are within 31.3 years of the mean age.
   d. None of these.

6. The top nine scores on the organic chemistry midterm are as follows: 74, 39, 80, 57, 43, 69, 21, 34, 69.
   Christine is currently taking college astronomy. The instructor often gives quizzes and the following are the top seven scores on the last quiz: 44 20 37 28 19 52 55.
   Which set of scores is shows more variation and why?

7. Below is the standard deviation for extreme 10K finish times for a randomly selected group of women and men. Which statement best summarizes the meaning of standard deviation?

   Women: $s = 0.16$ Men: $s = 0.25$
   e. The distribution of women's finish times is less varied than the distribution of men's finish times.
   f. The distribution of men's finish times is less varied than the distribution of women's finish times.
   g. On average, men's finish times will be 0.25 hours faster than the overall average finish time.
   h. On average, women's finish times will be 0.16 hours less than men's finish times.

8. Which of the following measurements is likely to have the **most** variation?
   i. The volume of individual pop cans measured in fluid ounces from a randomly selected twenty-four pack.
   j. The individual weights in ounces of tennis balls in a randomly selected can of tennis balls.
   k. The individual weights in ounces of potatoes in a randomly selected crate of potatoes.

Chapter 3 37

Table Number:_____   Group Name: _____

Group Members:_____   _____   _____

What makes the standard deviation larger or smaller?

Part I

Before we begin the attached worksheet, we are going to think about the meaning of typical, or standard, deviation from the mean. First, examine the following dot plot which has the mean marked in the plot. Think about how large the deviations would be for each data point (dot).

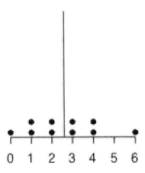

Next we will draw in each deviation from the mean.

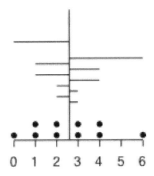

Now, think about the average size (length) of all of those deviations, and use this as an estimate for the size of the Standard Deviation. Don't worry about whether the deviation is to the left or right of the mean. Just consider all of the lengths. Draw the length of the standard deviation below.

Based on the scale in the graph, estimate a numerical value for the length of the line you drew above.

Repeat the process with this dot plot to help you draw and estimate the length of the standard deviation.

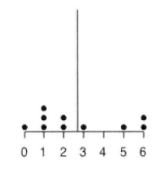

Now, try to draw and estimate the length of the standard deviation with the following histogram. The mean of the scores is 2.5. (Hint: Sketch in the appropriate number of dots in each bar of the histogram to make sure you have the appropriate number of deviations.)

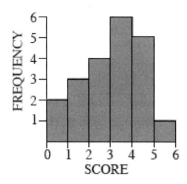

Comparing Standard Deviations

Below, you will find five pairs of graphs. The mean for each graph μ is given just above each histogram. For each pair of graphs presented,

- Indicate which one of the graphs has a larger standard deviation or if the two graphs have the same standard deviation.

- Explain why. (Hint: Try to identify the characteristics of the graphs that make the standard deviation larger or smaller.)

1.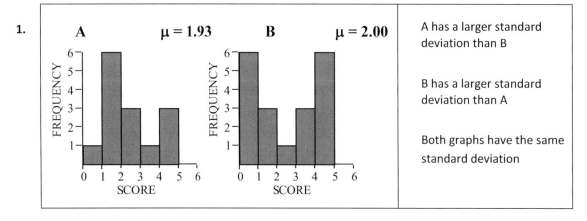

A has a larger standard deviation than B

B has a larger standard deviation than A

Both graphs have the same standard deviation

Explain.

2.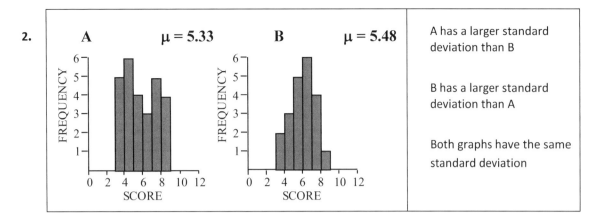

Explain.

3.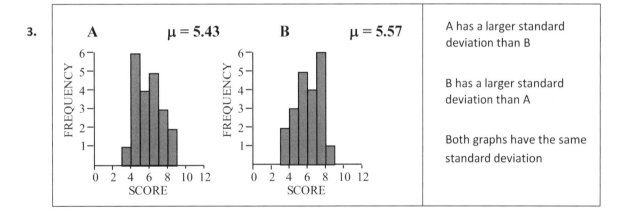

Explain.

4.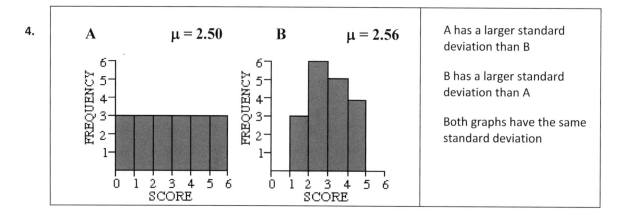

- A has a larger standard deviation than B
- B has a larger standard deviation than A
- Both graphs have the same standard deviation

Explain.

5.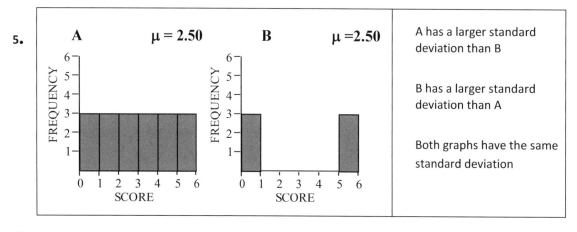

- A has a larger standard deviation than B
- B has a larger standard deviation than A
- Both graphs have the same standard deviation

Explain.

Reference

delMas, R.C. (2001b). What makes the standard deviation larger or smaller? STAR library. Retrieved October 21, 2007, from http://www.causeweb.org/repository/StarLibrary/activities/delmas2001/

Table Number:_____ Group Name: _____

Group Members:_____ _____ _____

Standard Deviation – Applied Problems

1. The data below are the lifetimes (in hours) for 10 light bulbs from a new brand that our school is considering for use in the football stadium light fixtures:

   2009 2015 2002 1979 2032 1991 2016 2001 1900 2055

   a) What is the value of the standard deviation of the lifetimes for the 10 light bulbs from the new brand? Show your calculations below.

   b) The standard deviation for the lifetimes of the bulbs from the brand currently in use is 40 hours. What does the standard deviation that you computed for the sample of light bulbs from the new brand tell you about how this brand might compare with the old brand?

   c) Replacing stadium light bulbs is difficult and requires special equipment. Because of this, rather than replace individual bulbs as they burn out, the school plans to replace *all* the stadium light bulbs as soon as one burns out. The mean lifetime is 2000 hours for both the current brand and the new brand under consideration, and the cost of the two brands is the same. Would you recommend that the school stay with the current brand or change to the new brand? Explain your reasoning.

2. A class of students took a test in Language Arts. The teacher determines that the mean grade on the exam is a 65%. She is concerned that this is very low, so she determines the standard deviation to see if it seems that most students scored close to the mean, or not. The teacher finds that the standard deviation is high. Interpret what might have caused this high standard deviation. Be detailed, but concise.

3. An employer wants to determine if the salaries in one department seem fair for all employees, or if there is a great disparity. He finds the average of the salaries in that department and then calculates the variance, and then the standard deviation. The employer finds that the standard deviation is slightly higher than he expected. Explain in some detail what might be causing the high standard deviation.

4. Consider the dotplots below showing the average temperature for the month of February 2006 for three cities. Which city would you expect to have the largest standard deviation in temperatures for the month? Explain how having this information might affect your choice of cities to which to move if given the opportunity. Be detailed, but concise.

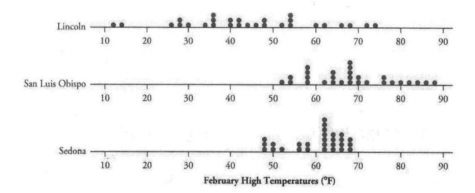

Are Female Hurricanes Deadlier than Male Hurricanes?

Background (Adapted from: "Female Hurricanes are Deadlier than Male Hurricanes, Study Says." by Holly Yan, CNN. June 3, 2014: http://www.cnn.com/2014/06/03/us/female-hurricanes-deadlier/)

Apparently sexism isn't just a social problem -- if you're in the path of a hurricane, gender bias might actually kill you. A study suggests people prepare differently for hurricanes depending on whether the storm has a male or female name. "Feminine-named hurricanes (vs. masculine-named hurricanes) cause significantly more deaths, apparently because they lead to a lower perceived risk and consequently less preparedness," a team of researchers wrote in the Proceedings of the National Academy of Sciences. In other words, a hurricane named "Priscilla" might not make people flee like a hurricane named "Bruno" would.

The study analyzed death rates from U.S. hurricanes from 1950 to 2012. "For severe storms, where taking protective action would have the greatest potential to save lives, the masculinity-femininity of a hurricane's name predicted its death toll," the study said. Hurricane Katrina in 2005, which left more than 1,800 people dead, was not included in the study because it was considered a statistical outlier. Neither was Hurricane Audrey in 1957, which killed 416 people. The study does note that both of those very deadly hurricanes had female names.

Why name hurricanes anyway? Giving hurricanes short, easy-to-remember names helps reduce confusion when two or more tropical storms are brewing at the same time, the National Hurricane Center said. For decades, all hurricanes were given female names in part because hurricanes were unpredictable, the study said, citing the "Encyclopedia of Hurricanes, Typhoons and Cyclones." "This practice came to an end in 1979 with increasing societal awareness of sexism, and an alternating male-female naming system was adopted," the report said.

Each year's list of hurricane names is alphabetical, alternating between male and female monikers. A U.N. World Meteorological Organization committee has already set up six years' worth of names. The lists repeat after each six-year cycle. "The only time that there is a change is if a storm is so deadly or costly that the future use of its name on a different storm would be inappropriate for obvious reasons of sensitivity," the National Hurricane Center said.

The hurricane data is given in an Excel Spreadsheet in the chapter 3 folder on Blackboard. It is also in our class data on Stat Crunch. It gives data that was used in the article *Female Hurricanes are Deadlier than Male Hurricanes* by Kiju Junga, Sharon Shavitta, Madhu Viswanathana, and Joseph M. Hil. In *Proceedings of the National Academy of Sciences of the United States of America*, May 2014.

*Note: hurricanes Katrina in 2005 (1833 deaths) and Audrey in 1957 (416 deaths) were removed from the data set.

Table Number:_____   Group Name: _____

Group Members:_____   _____   _____

Are Female Hurricanes Deadlier than Male Hurricanes?

1. Suggest a graph that might be used to compare the death totals for Female and Male named hurricanes. Explain why you chose the graph that you did.

2. Calculate the mean, standard deviation, and five-number summary of the death totals for Female and Male named hurricanes.

Gender	Mean	S.D.	Min	Q1	Median	Q3	Max
Female							
Male							

(a) Which measure, the mean or the median, do you think better represents a typical number of deaths from a hurricane? Why?

(b) Based upon the numerical calculations, do you think that the Female named hurricanes are more deadly? Why or why not?

3. For each of Female and Male named hurricanes, determine whether there are any outliers.

4. Construct comparative boxplots on Stat Crunch that display the distributions of the number of deaths for Female and Male named hurricanes. Copy and paste your boxplots below. Be sure to label your graphs' axes.

5. Thoroughly interpret the boxplots. Compare and contrast center and spread for the two distributions. Then, state your opinion on whether or not it seems that the Female named hurricanes are more severe.

6. How could the fact that all hurricanes had female names until 1979 bias the results from Question 5?

7. Now, consider only the *Female* named hurricanes. Earlier, it was noted that hurricanes Audrey and Katrina were omitted from the analysis. Add the death totals from these two hurricanes to your dataset and redo the summary calculations:

Katrina/Audrey Included	Mean	S.D.	Min	Q1	Median	Q3	Max
No							
Yes							

Which measure, the mean or the median, do you think better represents a typical number of deaths from a hurricane? Why?

Chapter 3 47

Table Number:_____   Group Name: _____

Group Members:_____   _____   _____

Interpreting Boxplots

Part I

The following graph shows the distribution of ages for 72 recent Academy Award winners split up by gender (36 females and 36 males).

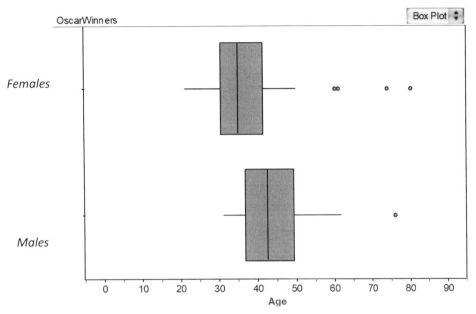

Use the graph to help answer the following questions.

a) Estimate the percentage of female Oscar winners who are younger than 40.

b) The oldest 50% of Oscar winners who are male are between which two ages?

c) What is the shape of the distribution of male Oscar winners? Explain.

d) Explain how to find the Inter-Quartile Range (IQR) for the female Oscar winners.

e) Find the IQR for the female Oscar winners.

f) What information does the IQR of the female Oscar winners offer us? Why would a statistician be more interested in the IQR than in the range?

g) Compare the medians for male and female Oscar winners. What do you conclude about the ages of male and female Oscar winners? Explain.

h) Compare the IQR for the male and female Oscar winners. What do you conclude about the ages of male and female Oscar winners now? Explain.

Part II

In the next problem, you will be given a descriptive scenario and a graph that shows two box plots. Use the graphs to make an informed comparison of the groups.

- Be sure to compare *shape*, *center* and *spread* of the distributions.
- Also, be sure that you are comparing the groups using the context of the data and not just comparing two (or more) numbers.

Stephen wants to investigate differences in spending habits of males and females. He compares the amounts spent per week on reading materials by males and females in a random sample of college students by generating the following plots.

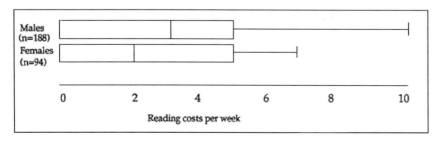

Help Stephen by comparing the two distributions in the space below.

Reference
Garfield, J., Zieffler, A., & Lane-Getaz, S. (2005). EPSY 3264 Course Packet, University of Minnesota, Minneapolis, MN.

Chapter 3 **51**

Table Number:_____ Group Name: _____

Group Members:_____ _____ _____

Matching Histograms to Boxplots

Consider all of the graphs below, the histograms and the boxplots, to be on the same scale. Match each histogram with its corresponding boxplot, by writing the letter of the box plot in the space provided.

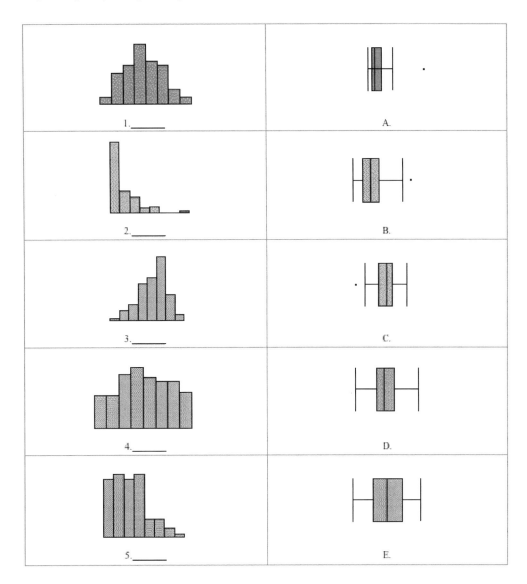

Reference

Scheaffer, R.L., Watkins, A., Witmer, J., & Gnanadesikan, M. (2004a). *Activity-based statistics: Instructor resources* (2nd edition, Revised by Tim Erickson). Key College Publishing.

Chapter 4

Regression Analysis: Exploring Associations between Variables

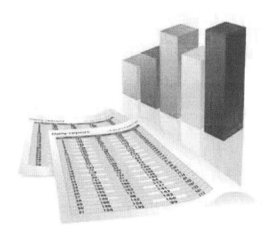

Table Number:_____   Group Name: _____

Group Members:_____   _____   _____

Scatter Plots

The **scatter plot** is the basic tool used to investigate relationships between two **numeric or quantitative** variables.

What do you see in these scatter plots? Write a description for each which includes **trend, shape,** and **strength** and explain what all these mean **in the context of the data**. When describing the trend, use the words "increasing" or "decreasing." Describe the shape as being linear or non-linear. When describing strength, make note of how close together the points are. Use one sentence to interpret the graph in context.

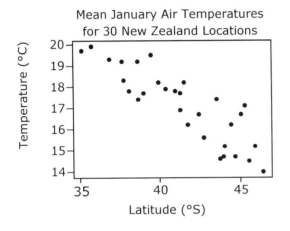

1. **Trend:** _____

   Shape: _____

   Strength: _____

   Interpret: _____

   _____

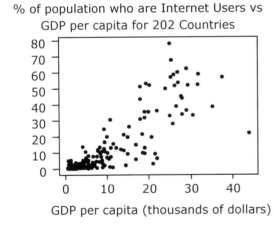

2. **Trend:** _____

   Shape: _____

   Strength: _____

   Interpret: _____

   _____

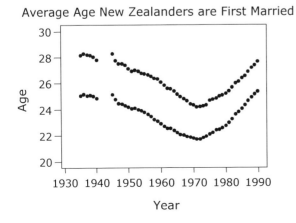

3. **Trend:** _____

   Shape: _____

   Strength: _____

   Interpret: _____

   _____

4. Rank these relationships from weakest to strongest:

A)
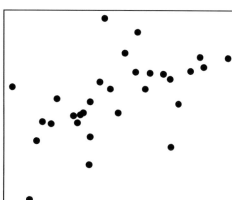

B)
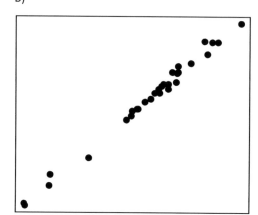

C)
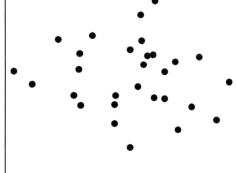

D)
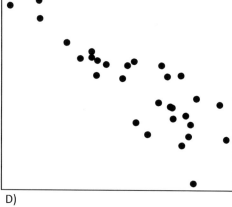

Write your ranking, from weakest to strongest here, using the letters that represent the graphs:
_____

Explain your reasoning please!

_____
_____
_____

5. In the following scatterplot, circle the data point that might be an outlier and list its coordinates here: _____ . Interpret the coordinates of this point in context.
   _____

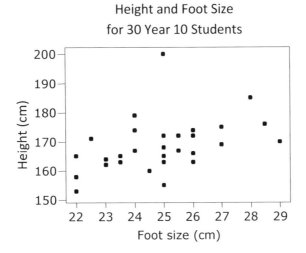

What will happen to the correlation coefficient if the outlier is removed?
(Remember the correlation coefficient answers the question: "For a linear relationship, how well do the data fall on a straight line?")

☐ It will get smaller   ☐ It won't change   ☐ It will get bigger

6. **What will happen to the correlation coefficient if the Elephant data point is removed?**
(Remember the correlation coefficient answers the question: "For a linear relationship, how well do the data fall on a straight line?")

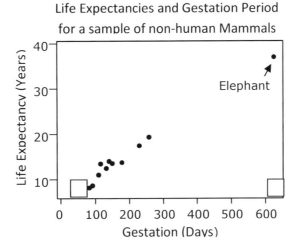

☐ It will get smaller
☐ It won't change much
☐ It will get bigger

Table Number:_____   Group Name: _____

Group Members:_____   _____   _____

Lurking Variables

For questions #1 and #2, follow the directions given.

1.

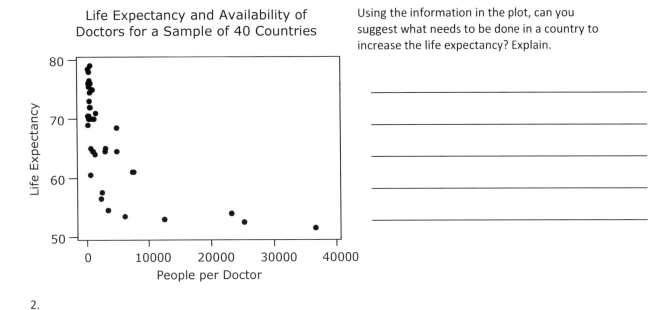

Life Expectancy and Availability of Doctors for a Sample of 40 Countries

Using the information in the plot, can you suggest what needs to be done in a country to increase the life expectancy? Explain.

_____
_____
_____
_____
_____

2.

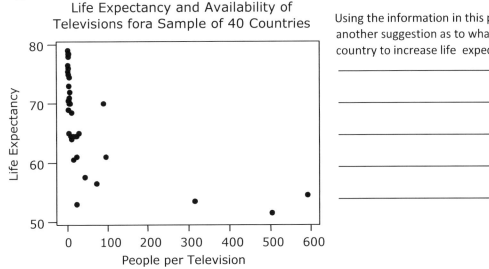

Life Expectancy and Availability of Televisions for a Sample of 40 Countries

Using the information in this plot, can you make another suggestion as to what needs to be done in a country to increase life expectancy?

_____
_____
_____
_____
_____

Can you suggest another variable that is linked to life expectancy and the availability of doctors (and televisions) which explains the association between the life expectancy and the availability of doctors (and televisions)? We call this type of variable a *lurking variable.*

For each of the scenarios described below, list at least one possible lurking variable.

1. FIRE DAMAGE AND FIRE TRUCKS
   The amount of damage at the scene of a fire has a positive correlation with the number of fire trucks on the scene.

2. BUNDLED-UP BABIES AND OUTDOOR TEMPERATURE
   The age at which a baby first learns to crawl has a negative correlation with outdoor temperature.

3. DANGEROUS ICE CREAM AND DROWNINGS
   The number of ice cream servings consumed has a positive correlation with the number of drownings.

4. CAR COST AND GAS MILEAGE
   In 2005, there was a negative correlation between car cost and gas mileage.

5. FOOT SIZE AND READING ABILITY
   Foot size and reading ability have a positive correlation.

Diamond Rings

The following scatterplot contains the prices of ladies' diamond rings and the carat size of their diamond stones. All of the rings are made with gold of 20 carats purity and are each mounted with a single diamond stone. A line of best fit has also been added to the scatterplot. The equation for this line is as follows:

$$\hat{\$} = -154.94 + 2220.66(Carat)$$

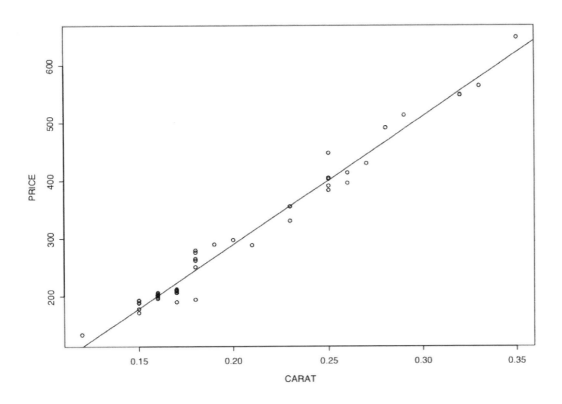

1. Does the line of best fit seem like a good model for the data? Explain.

2. What is the *slope* for the line of best fit depicted in the scatterplot?

3. What does the slope tell us? (*Hint: Interpret the slope in the "real world" context of the data.*)

4. What is the *y-intercept* for the line of best fit depicted in the scatterplot?

5. What does the y-intercept tell us? *(Hint: Interpret the y-intercept in the "real world" context of the data.)*

6. Does this y-intercept make sense given your interpretation in Question 5? Explain.

Reference

Chu, S. (2001, July). Pricing the C's of diamond stones. *Journal of Statistics Education, 9*(2). Retrieved December 6, 2006, from http://www.amstat.org/publications/jse/v9n2/datasets.chu.html

Table Number:_____   Group Name:_____

Group Members:_____   _____   _____

Activity: The Regression Line

The data set, which is stored in the file North Carolina birth data, is a random sample of 1000 birth records taken by the North Carolina State Center for Health and Environmental Statistics in 2005. Of particular interest will be incidents of low infant birth weight. Low birth weight is defined as less than 2500 grams. Then goal of this activity will be to summarize the variables in this data set both graphically and numerically. The variables in the study are:

Variable Label	Description of variable
Plurality	Refers to the number of children associated with the birth
Sex	Sex of child (Gender 1=Male 2=Female)
Fage	Age of father (years)
Mage	Age of mother (years)
Weeks	Completed weeks of gestation
Visits	Number of pre-natal medical visits
Marital	Marital status (1=married 2=unmarried)
Racemom	Race of Mother (0=Other Non-white 1=White 2=Black 3=American Indian 4=Chinese 5=Japanese 6=Hawaiian 7=Filipino 8=Other Asian or Pacific Islander)
Hispmom	Whether mother is of Hispanic origin (C=Cuban M=Mexican N=Non-Hispanic O=Other and Unknown Hispanic P=Puerto Rican S=Central/South American U=Not Classifiable)
Gained	Weight gain during pregnancy (pounds)
Lowbw	If birth weight is 2500 grams or lower, 0=infant was not low birth weight, 1=infant was low birth weight
Tpounds	Birth weight in pounds
Smoke	0=no 1=yes for mother admitted to smoking
Mature	0=no for mother is 34 or younger 1-yes for mother is 35 or older
Premie	0=no 1=yes to being born 36 weeks or sooner.

1. Answer the following for the variables Tpounds (response variable, y) regressed on Weeks (Explanatory Variable, x).
    A. Make a scatterplot of the data with the regression line. Report the parameter estimates (estimates of the slope and intercept). Copy and paste your information here:

B. Interpret the slope and the intercept

C. Use the coefficient of determination to determine the percentage of the variation in Tpounds that is explained by Weeks.

D. What is the predicted value for Tpounds when Weeks is 35? What if Weeks is 40?

E. How well does the model fit the data? How do you know?

2. Answer the following for the Fage (response variable) regressed on Mage (Explanatory Variable).
   A. Make a scatterplot of the data with the regression line. Report the parameter estimates (estimates of the slope and intercept). Copy and paste your information from StatCrunch here:

B. Interpret the slope and the intercept in context of the "real world" data.

C. Use the coefficient of determination to determine the percentage of the variation in Fage that is explained by Mage?

D. What is the predicted value for Fage when Mage is 35? What if Mage is 17?

E. How well does the model fit the data? How do you know?

Answer the following using the information calculated

3. A friend of yours is pregnant with her first child. She is nervous that she might have her baby at 34 weeks. What can you tell her about the predicted weight of the baby?

4. Your cousin is 40 years old. Predict the age of the father of the child. Using the scatterplot, give a range for the suspected ages of the father.

5. Of the two regression models, which has the best fit and why?

If time permits try one more set:

Answer the following for the Tpounds (response variable, y) regressed on Mage (Explanatory Variable, x).

    A. Make a scatterplot of the data with the regression line. Report the parameter estimates (estimates of the slope and intercept)

    B. Interpret the slope and the intercept

    C. Use the coefficient of determination to determine the percentage of the variation in Tpounds that is explained by Mage?

    D. What is the predicted value for Tpounds when Mage is 35? What if Mage is 17?

Table Number:_____   Group Name: _____

Group Members: _____   _____   _____

Text Messaging is Time Consuming! What Gives?

Introduction

Many of us send lots of text messages throughout a day. What factors could be related to the number of text messages one sends in a day? In this activity, we will explore the relationship between the number of text messages one sends in a day and a few other potential explanatory factors.

With your group, choose one of the following questions to explore:
- A. Does the number of hours you spend hanging out with friends in a day increase or decrease with the number of text messages you send?
- B. Does the number of hours you spend doing homework in a day increase or decrease with the number of text messages you send?
- C. Does the number of text messages you receive in a day increase or decrease with the number of text messages you send?

To answer the question you choose, you are going to download and work with a real data set. To download the data set, go to the following website: http://www.amstat.org/censusatschool/

With your group, please carry out the following steps:

a) Click on Random Sampler

b) Accept the Terms & Conditions

c) Select a sample size of 100 from All States and 9, 10, 11, and 12 grade levels. Include All Genders and All Years of data collection.

d) Download the data into Excel.

e) Open the data in Excel. You will see a large number of variables (labeled in each column).

f) Delete all the columns except for the following:

Gender, Text Messages Sent Yesterday, Text Messages Received Yesterday, Hanging out with Friends Hours, Doing Homework Hours

g) Now open StatCrunch and copy and paste these remaining columns into StatCrunch

****

1. Determine which of your variables is the **dependent (or response or predicted) variable** (y) and which is the **independent (or explanatory or predictor) variable** (x) and write them below.

    Dependent (or response or predicted) variable: _____

    Independent (explanatory or predictor) variable: _____

2. Create a scatterplot with StatCrunch and copy and paste it below then answer the given questions.

   i. Does the relationship appear to be linear? Explain.

   ii. Are there any outliers? If so, list them here:_____

   iii. What are some possible explanations for why there could be outliers?

3. Use StatCrunch to estimate the least squares regression line for your downloaded data and write your equation here:

   i. What is the slope of your equation?_____ Interpret this slope in the context of the problem

   ii. What is the y-intercept of your equation? _____ Interpret this y-intercept in the context of the problem.

   iii. What is the correlation coefficient? _____ Interpret this in the context of the problem.

   iv. What percentage of the variation in your response variable is explained by the explanatory variable? Explain.

Now choose a DIFFERENT question A – C to answer and repeat steps 1 – 3.

1. Determine which of your variables is the **dependent (or response or predicted) variable** (y) and which is the **independent (or explanatory or predictor) variable** (x) and write them below.

    Dependent (or response or predicted) variable: _____

    Independent (explanatory or predictor) variable: _____

2. Create a scatterplot with StatCrunch and copy and paste it below then answer the given questions.

    i. Does the relationship appear to be linear? Explain.

    ii. Are there any outliers? If so, list them here:_____

    iii. What are some possible explanations for why there could be outliers?

3. Use StatCrunch to estimate the least squares regression line for your downloaded data and write your equation here:

    i. What is the slope of your equation?_____ Interpret this slope in the context of the problem.

    ii. What is the y-intercept of your equation? _____ Interpret this y-intercept in the context of the problem.

    iii. What is the correlation coefficient? _____ Interpret this in the context of the problem.

    iv. What percentage of the variation in your response variable is explained by the explanatory variable? Explain.

Written by
Jeanie Gibson, Mary McNelis, and Anna Bargagliotti (for Project-SET)
Hutchison School, St. Agnes Academy, and Loyola Marymount University

Table Number:_____   Group Name: _____

Group Members:_____   _____   _____

Practice Interpreting Slopes and Y-Intercepts

1. Suppose a market researcher is interested in studying the relationship between the odometer reading on a used car and its selling price. He collects a random sample of 100 cars and uses technology to create the following regression equation: **Selling Price = 6,500 – 0.0312(miles on odometer).**

    a) Name the slope and interpret it in the context of the problem. Use a complete sentence please!

    b) Name the y-intercept and interpret it in the context of the problem. Does your interpretation make sense? Is 0 a reasonable value for the explanatory variable?

2. Given that the explanatory variable, x, is a worker's commute time in minutes and the response variable, y, is the score on a well-being survey. The least squares regression line is:
    Well being = -0.0479(commute time) + 69.0296.

    a) Name the slope and interpret it in the context of the problem. Use a complete sentence please!

    b) Predict the well-being index of a person whose commute time is 30 minutes.

3. A researcher wanted to know if cola consumption is associated with lower bone mineral density in women. The least squares regression line treating cola consumption as the explanatory variable (in number of colas per week) is: **Bone density = -0.0029(cola consumption) + 0.8861** where bone mineral density is measured in grams per square centimeter.

    a) Name the slope and interpret it in the context of the problem. Use a complete sentence please!

    b) Name the intercept and interpret it in the context of the problem. Does your interpretation make sense? Is 0 a reasonable value for the explanatory variable?

    c) Predict the bone mineral density of a woman who consumes four colas a week.

4. A researcher collected data representing the number of days absent, x, and the final grade, y, for a sample of college students in a general education course at a large state university. The regression equation is:
   Final grade = -2.8273(number of absences) + 88.7327.

   a) Name the slope and interpret it in the context of the problem. Use a complete sentence please!

   b) Name the intercept and interpret it in the context of the problem. Does your interpretation make sense? Is 0 a reasonable value for the explanatory variable?

   c) Predict the final grade for a student who misses five classes.

5. A researcher studied the relationship between the total number of touchdowns, x, made by top paid National Football League (NFL) quarterbacks and their salaries, y. The regression equation is:
   Salary = 4,393,649.84 + 320,510.26(number of touchdowns).

   a) Name the slope and interpret it in the context of the problem. Use a complete sentence please!

   b) Name the intercept and interpret it in the context of the problem. Does your interpretation make sense? Is 0 a reasonable value for the explanatory variable?

   c) If $r = 0.5954$ and $r^2 = 0.3545$, what can you say about the strength of the association between touchdowns scored and a quarterback's salary? How much of the variation in salary can be explained by the number of touchdowns scored?

Chapter 5

Modeling Variation with Probability

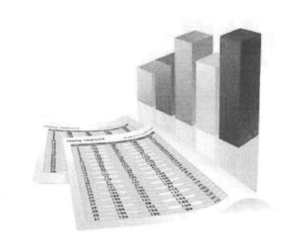

  # Random Babies

Suppose that on one night at a certain hospital, four mothers (named Smith, Jones, Marshall, and Banks) gave birth to baby boys. Suppose the babies names are Sam Smith, Joe Jones, Mark Marshall, and Bobby Banks. Suppose also that, as a very sick joke, the hospital staff decides to return babies to their mothers completely at random. (Totally fiction – no hospital staff we know would even think of doing this!)

The questions we want to investigate are these: How often will at least one mother get the right baby? How often will every mother get the right baby? What is the most likely number of correct matches of baby to mother? On average, how many mothers will get the right baby?

We will use a *simulation* to investigate what happens in the long run.

Write the definition of *simulation* here: _____

_____

To represent the process of distributing babies to mothers at random, you will shuffle and deal 3x5 cards (representing the babies) to regions on a sheet of paper (representing the mothers).

1. Use the four index cards I'll give you and one piece of your own notebook paper. On each index card, write one of the babies' first names. Divide your sheet of paper into four regions, writing one of the mothers' last names in each region. Shuffle the four index cards well, and deal them out randomly, placing one card face down on each region of your paper. Finally, turn the cards over to reveal which babies were randomly assigned to which mothers. Record how many mothers received their own baby. Jot this number down under "Trial 1" in the table below.

Trial	1	2	3	4	5
Number of matches					

2. Repeat the random "dealing" of babies to mothers a total of five times, recording in each case how many mothers received the correct babies. Record your counts in the above table.

3. Now, you're going to combine your results with all the other students at your table. Each student needs to make a chart in her notes like the one at the top of the next page. Assign one student at your table to be the manager. The manager will ask each student, "How many of your trials resulted in AT LEAST ONE match?" Each student will record these numbers in Column 1 in her chart. Once all nine students have announced their results each student will now complete her own chart. (Feel free to help each other.) Notice that we are going to keep adding on the number of trials, so that we keep a CUMULATIVE COUNT of the total number of trials and a CUMULATIVE COUNT of then number of trials with at least one match. In the last column, figure out the proportion of trials with at least one match. Once completed with all nine students at your table, your chart should have nine rows and should begin like the one below. (The last row in Columns 3 and 4 in all charts should have the exact same numbers in them.)

Student #	Column 1: Number of trials with at least one match	Column 2: Cumulative Number of Trials	Column 3: Cumulative Number of Trials with at Least one Match	Column 4: Cumulative Proportion of Trials with at Least One Match
1		5	(student #1's count)	(column 3 ÷ column 2)
2		10	(student 1's + student 2's count)	
3		15		
4		20		
5		25		
6		30		
7		35		
8		40		
9		45		

4. Construct a graph of the cumulative proportion of trials with at least one match vs. the cumulative number of trials.

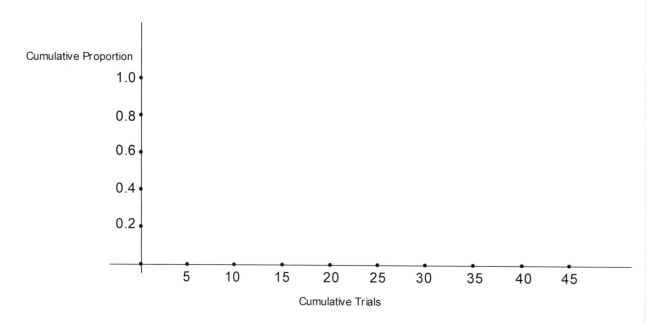

5. Does the proportion of trials that result in at least one mother getting the right baby fluctuate more at the beginning or at the end of this process?

6. Does the cumulative proportion (also known as the **cumulative relative frequency**) appear to be "settling down" and approaching one particular value? What do you think that value will be?

Complete the following:

The **probability** of a random event occurring is _____

_____

_____

_____

Simulation leads to an **empirical estimate** of the probability, which is _____

_____

Reference

Adapted from Activity 11-1: Random Babies in Rossman and Chance (2012), *Workshop Statistics, 4th ed*. John Wiley & Sons

Table Number:_____   Group Name: _____

Group Members:_____   _____   _____

Random Babies (Cont'd)

HAND IN THIS PAGE AND THE NEXT. WORK WITH YOUR GROUP AND HAND IN ONE COPY FOR THE GROUP. I WILL GRADE FOR ACCURACY AND NEATNESS. THE ASSIGNMENT IS WORTH 10 POINTS.

We can approximate these probabilities more accurately by performing more trials. Do the following with your group of three students.

1. Use the Random Babies applet to simulate this random process. You can find a link to the app in our Canvas course in the Chapter 5 module immediately after the "Random Babies Activities" heading in the In-Class Group Work section. Leave the number of trials at **1**, then click **Randomize** five times. Then turn off the **Animate** feature and ask for **995** more trials, for a total of 1000 trials.

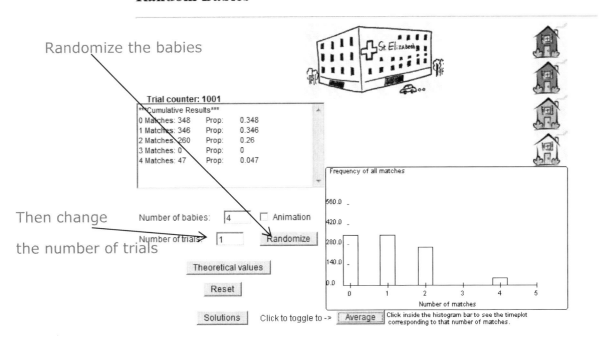

2. Record the counts and proportions in the following table

Number of matches	0	1	2	3	4	Total
Count						
Proportion (out of 1000)						1.00

3. Are these simulation results reasonably consistent with the results obtained at your table? Explain.

4. Report the new (likely more accurate) empirical estimate that at least one mother gets the correct baby.

5. Click the bar of the histogram corresponding to 0 matches to see a graph of how the relative frequency changes over time. Does this graph appear to be fluctuating less as more trials are performed, approaching a limiting value? Describe what you see.

6. Is an outcome of exactly 3 matches possible? Why or why not?

7. Is it impossible to get four matches? Is it likely?

Reference
Adapted from Activity 11-1: Random Babies in Rossman and Chance (2012), *Workshop Statistics, 4th ed*. John Wiley & Sons.

Chapter 5   81

Table Number:_____   Group Name: _____

Group Members:_____   _____   _____

Sample Spaces and Events

> **RECALL:**
> A **sample space**, S, of a probability experiment is the collection of all possible outcomes.
> An **event** is any collection of outcomes from aa probability experiment.

For each of the following probability experiments below, list the sample space and the indicated events.

1. EXPERIMENT: Rolling a single fair die
   a. Sample space: _____
   b. Event A: Rolling an even number. A= _____
   c. A^C = _____
   d. Event B: Rolling a number greater than three. B= _____
   e. B^C = _____
   f. Event C; Rolling at least a two. C= _____
   g. C^C = _____
   h. Event D: Rolling a number no more than 4. D= _____
   i. D^C = _____

2. EXPERIMENT: Rolling two fair die and consider the sum of the die.
   a. Sample space: (*The table below might help you list all the possible outcomes.*)

Die 1											
Die 2											

Die 1											
Die 2											

Die 1											
Die 2											

   b. Event A: Rolling a sum of 4:_____
   c. Event B: Rolling a sum of 10: _____
   d. Event C: Rolling a sum of 3:_____
   e. Event D: Rolling a sum greater than 4: _____
   f. D^C = _____

The following table might help you organize the above data for finding the requested events. List the sums of the two dice. (*Please complete.*)

	1	2	3	4	5	6
1						
2						
3						
4						
5						
6						

3. EXPERIMENT: A couple decides to have two children. Using "B" for boy and "G" for girl, list all the possible outcomes of boys and girls.

4. EXPERIMENT: A couple decides to have three children. Using "B" for boy and "G" for girl, list all the possible outcomes of boys and girls.

5. EXPERIMENT: Suppose you flip a coin 4 times. List all the possible outcomes of heads and tails.

6. EXPERIMENT: Three children, Alice, Bob, and Caren, are arguing about who should be first in line for the ride at Cedar Point. List all the possible ways they could line up. Let A = Alice, B = Bob, C = Caren.

Chapter 5   83

Table Number:_____   Group Name: _____

Group Members:_____   _____   _____

Basic Probability

1. If you draw one card randomly from a standard 52-card playing deck, what is the probability that it will be a:
    a. spade?

    b. red card?

    c. face card (jack, queen, or king?

    d. 7?

2. Suppose you roll a fair die one time. What is the probability of obtaining
    a. An even number?

    b. A number less than 4?

    c. A number no more than 3?

    d. A number that is at least a 2?

3. An exam consists of 20 multiple choice questions. Each of the answers is either right or wrong. List ALL the possible outcomes for each of the following events. List the number of wrong answers as an outcome.
    a. A student makes more than 11 mistakes

    b. A student makes no more than 11 mistakes

    c. A student makes at least 11 mistakes.

    d. Are any two of the above complementary? If so, which? Explain.

A *sample space* is a list of all possible (and equally likely) outcomes.

4. The sample space of an experiment where a fair coin is flipped twice is the following. Suppose H stands for heads and T stands for tails. The outcomes are equally likely.

$$HH, HT, TH, TT$$

What is the probability of getting:

a. Exactly 1 heads?

b. At least one heads?

c. No more than one head?

5. Write the sample space of all possible sequences for a family of three children. You should get 8 possible outcomes. The first is done for you.

G G G

a. What is the probability of having exactly 2 boys?

b. What is the probability of having all girls?

c. What is the probability of having at least one girl?

Chapter 5

Table Number: _____   Group Name: _____

Group Members: _____   _____   _____

A Sweet Task

Suppose that we counted the number of M&Ms and Skittles of each color in a bag of the respective candies and recorded the following data.

	Red	Orange	Yellow	Green	Blue	Brown	Total
M&Ms	12	6	8	14	10	8	58

	Red	Orange	Yellow	Green	Blue	Purple	Total
Skittles	9	11	7	10	7	5	49

1. **Create a Two-Way Frequency Table**: We can combine individual frequency tables into a two-way frequency table. The rows represent the types of candy and the columns represent the color of the candy. Use the data above to fill in the two-way frequency table below. Be sure to total each column and row.

	Red	Orange	Yellow	Green	Blue	Purple	Brown	Total
M&Ms								
Skittles								
Total								

We read a two-way frequency table in a similar way as a regular frequency table. For example, the number of orange Skittles is listed where the "Orange" column and the "Skittles" row meet. This is called a **joint frequency**.

We can also find the total number of blue candies in the bag. We just look at the total of the "Blue" column. This is a **marginal frequency**.

2. **Analyzing the Data – Finding Marginal and Joint Probabilities:** We can compute the probability of an event occurring from the frequency counts in the candy "mix" two-way frequency table. Find the probability of randomly choosing a candy from the "mix" with the listed attributes. Also, identify each event as either a joint or marginal probability. A joint probability requires two or more characteristics to hold true, whereas a marginal probability requires only one.

		Probability	Joint or Marginal Probability?
a.	Any Color M&M		
b.	A Purple Skittle		
c.	A Blue M&M		
d.	An Orange Skittle		
e.	Any Green candy		
f.	A Blue Skittle		

g. In your own words describe how you compute a joint probability given counts in a two-way frequency table.

h. In your own words describe how you compute a marginal probability given counts in a two-way frequency table.

3. **Finding Conditional Probability with Counts:** Imagine that your friend chooses a candy piece from the above "mix". She looks at it, tells you that it is red, but doesn't tell you if it is an M&M or a Skittle.

Knowing that your friend has a red candy in her hand, we can find the probability that it is a red M&M. This is called the **conditional probability** of an event because **we already know something (a condition) about the event in question.**

Answer the following questions to help you find the conditional probability.

a. What is the "total number of <u>possible</u> outcomes" for your friend's candy? *(Remember we know the candy is red.)*

b. What is the probability that your friend has an M&M, if we know the candy is red? *(Keep in mind we only are worried about M&Ms that are red.)*

c. In your own words, explain how to compute conditional probabilities given a two-way frequency table.

> What you just found can be written as P(M&M | red), which we read as "the probability of a candy being an M&M *given* that it is red".

4. **Computing Conditional Probabilities:** Using the data about the candy "mix", find the following conditional probabilities. Please show your set up and then your answer as either a simplified fraction or a decimal

   a. P (green | Skittle) _____   b. P (M&M | blue) _____

   c. P (brown | M&M) _____   d. P (Skittle | red) _____

   e. P (Skittle | purple) _____   f. P (M&M | purple) _____

   g. P(yellow | M&M) _____   h. P(M&M | yellow) _____

5. If you draw a red candy, is it more likely to be an M&M or Skittle? _____ Why?

Table Number:_____   Group Name: _____

Group Members:_____   _____   _____

Using "AND" or "OR" to Combine Events

1. Kent State conducted a survey of 277 sophomore, junior, and senior undergraduate students regarding satisfaction with their living quarters. Results of the survey are shown in the table by class rank.

	Sophomore	Junior	Senior	Total
Satisfied	46	67	62	175
Neutral	14	17	10	41
Not satisfied	20	15	26	61
Total	80	99	98	277

A survey participant is selected at random. What is the probability that she is…

a) A sophomore? In other words, what is *P(sophomore)*?

b) Neutral? In other words, what is *P(neutral)*?

c) A sophomore AND neutral? In other words, what is *P(soph AND neutral)* ?

d) A junior AND not satisfied? In other words, what is *P(junior AND not satisfied)* ?

e) A junior OR not satisfied ? In other words, find *P(junior OR not satisfied)*.

f) *P(senior or neutral)?*

g) *P(senior or not satisfied)?*

2. The following data represent the number of drivers involved in fatal crashes in the United States in 2009 by day of the week and gender.

	Male	Female	Total
Sunday	8,222	4,325	**12,547**
Monday	6,046	3,108	**9,154**
Tuesday	5,716	3,076	**8,792**
Wednesday	5,782	3,011	**8,793**
Thursday	6,315	3,302	**9,617**
Friday	7,932	4,113	**12,045**
Saturday	9,558	4,824	**14,382**
Total	**49,571**	**25759**	**75,330**

a) Determine the probability that a randomly selected fatal crash involved a male.

b) Determine the probability that a randomly selected fatal crash occurred on a Sunday.

c) Determine the probability that a randomly selected fatal crash occurred on a Sunday and involved a male.

d) Determine the probability that a randomly selected fatal crash occurred on Sunday or involved a male.

e) Would it be unusual for a fatality to occur on Wednesday and involve a female driver?

Table Number:_____   Group Name: _____

Group Members:_____   _____   _____

CONDITIONAL PROBABILITY APP ACTIVITY

Go to   http://onlinestatbook.com/2/probability/conditional_demo.html

Instructions

The simulation shows a set of 30 objects varying in color (red, blue, and purple) and shape (X and O). One of the objects is to be chosen at random. The various possible conditional probabilities are shown below the objects. Calculate each probability by counting the appropriate objects. Check your work by clicking on the radio button to the left. For example, if you click on P(X|Red) which is read "the probability of X given Red" then a box is put around each of the red objects. Of these, those that are X's are shaded. The probability of X given it is red is the number of red X's (shaded boxes) divided by the total red items (the number of boxes). Do this several times until you feel comfortable using the app to compute conditional probabilities. Click "Another Example" for new distribution of objects.

Once you are comfortable with finding conditional probabilities, please use the set of Xs and 0s below to determine the conditional probabilities on the following page.

O	X	X	X	X	O
O	X	O	O	X	O
X	O	O	O	O	X
X	X	X	O	O	O
O	O	X	O	X	X

[Another Example]

Please click on one of the radio buttons to see the conditional probability:

○ P(X|Red)   ○ P(Red|X)

○ P(X|Blue)   ○ P(Blue|X)

○ P(X|Purple)   ○ P(Purple|X)

○ P(O|Red)   ○ P(Red|O)

○ P(O|Blue)   ○ P(Blue|O)

○ P(O|Purple)   ○ P(Purple|O)

1. $P(X|\text{Red})$

2. $P(X|\text{Blue})$

3. $P(X|\text{Purple})$

4. $P(O|\text{Red})$

5. $P(O|\text{Blue})$

6. $P(O|\text{Purple})$

7. $P(\text{Red}|X)$

8. $P(\text{Blue}|X)$

9. $P(\text{Purple}|X)$

10. $P(\text{Red}|O)$

11. $P(\text{Blue}|O)$

12. $P(\text{Purple}|O)$

Table Number:_____   Group Name: _____

Group Members:_____   _____   _____

Writing conditional probabilities symbolically

Write each of the following probabilities symbolically:

1. The probability that a randomly selected teenager (12-17 years old) from the United States will go online at some point during the month is 93%. Event A: the selected person is a teenager. Event B: the selected person goes online during the month.

2. The probability that a randomly selected adult age 65 or older will go online during the month is 38%. Event A: the selected person is 65 or older. Event B: the selected person will go online during the month.

3. The probability that a randomly selected resident of the United States is a teenager who will go online this month is about 7%. Event A: the selected person is a teenager. Event B: the person will go online this month.

Table Number:_____   Group Name: _____

Group Members:_____   _____   _____

Mutually Exclusive Events and the Addition Rule

A probability experiment is conducted in which the sample space of the experiment is
$S = \{1,2,3,4,5,6,7,8,9,10,11,12\}$. Let the event $E = \{2,3,4,5,6,7,\}$, event $F = \{5,6,7,8,9\}$, event $G = \{9,10,11,12\}$, and event $H = \{2,3,4\}$. Assume that each outcome is equally likely.

1. List the outcomes in E AND F. _____
   Are E and F mutually exclusive? Explain.

2. List the outcomes in F AND G. _____
   Are F and G mutually exclusive? Explain

3. List the outcomes in F or G. _____
   Now find $P(F \text{ or } G)$ in two ways:
   a) by counting the number of outcomes in F or G;

   b) by using Rule 4.

4. List the outcomes in E or H. _____
   Now find $P(E \text{ or } H)$ in two ways:
   a) by counting the number of outcomes in E or H

   b) by using Rule 4

5. List the outcomes in E AND G. _____

   a) Are E and G mutually exclusive? _____ Explain.

   b) Find $P(E \text{ or } G)$ in two ways:

   by counting the outcomes in E or G

   by using the addition rule

6. List the outcomes in E^c : _____

   Find $P(E^c)$

Table Number:_____   Group Name: _____

Group Members:_____ _____ _____

SIMULATION ACTIVITY

To prepare: Each individual chooses
- 1 ace
- 2 face cards
- 3 number cards

A candy company is having a contest. Each candy bar wrapper has one letter printed on its inside. The letters are W, I, and N and they are printed in ratios of 3:2:1 respectively. If you spell the word "WIN" with candy bar wrappers, you receive a year's supply of the candy bars. To determine how many candy bars you should buy to spell WIN, perform the following simulation:

1. Let a number card represent a *W*, a face card represent an *I*, and an ace represent an *N*.
2. Shuffle the cards. Choose one card and record the corresponding letter below.

3. Repeat the process of shuffling, choosing, and recording until each letter has been obtained.

4. Record the number of tries it took you to spell out the word "WIN."

5. Repeat this simulation two more times and record how many times you had to shuffle the cards and choose one before you spelled WIN.

6. Combine your results with others at your table so that you have a total of 27 trials. From this combined data, estimate a reasonable number of candy bars that need to be purchased in order to WIN. Explain your reasoning. How did you come up with this estimate?

Chapter 5   99

Table Number:_____   Group Name: _____

Group Members:_____   _____   _____

Modeling a Random Phenomenon
Matching Chocolates

A bowl of chocolate candies has the following colors: 6 gold
                                                            8 blue
   7 green
                                                            4 brown

What is the probability of getting two of the same color candies when randomly selecting two chocolate candies from the bowl?

Simulation:
1. Describe a simulation that could be used to answer this question.

2. Perform at least 30 trials of the simulation at your table. Record the results below. Let G = gold; B = blue; R = green; N = brown.

3. Interpret your results as you answer the question: what is the probability of getting two of the same color candies when randomly selecting two chocolate candies from the bowl?

Theoretical Probability:
4. Describe the sample space that would mean getting two candies of the same color. Write all possible outcomes below.

5. Draw a probability tree diagram for the scenario on the back of this sheet.

6. Use the sample space and tree diagram to answer the question: what is the (theoretical) probability of getting two of the same color candies when randomly selecting two chocolate candies?

Compare your answers for parts (c) and (f).

Table Number:_____   Group Name: _____

Group Members:_____   _____   _____

Independent Events (and other stuff) – Skill Builder

Remember the test for independent events: TWO EVENTS ARE INDEPENDENT WHENEVER $P(B|A)=P(B)$

1. Suppose that for events A and B, $P(A)=0.4$ and $P(B)=0.3$, and $P(A \text{ and } B)=0.1$

    a) Are A and B mutually exclusive?

    b) Are A and B independent?

    c) Find P(A or B)

    d) Find $P(B^c)$

    e) Find $P(A|B)$

    f) Find $P(B|A)$

2. Suppose that for events A and B, $P(A)=0.8$, $P(B)=0.4$, and $P(A \text{ and } B)=0.25$.

    a) Are A and B mutually exclusive?

    b) Are A and B independent?

    c) Find P(A or B)

    d) Find $P(B^c)$

    e) Find $P(A|B)$

    f) Find $P(B|A)$

3. Suppose that events A and B are independent. Suppose also that $P(A)=0.7$ and $P(B)=0.6$. Find $P(A \text{ and } B)$

4. Determine if the two events (A and B) described are **mutually exclusive, independent,** and/or **complements.** (It's possible that the two events fall into more than one of the three categories or none of them.)

    Roll two (six-sided) dice. Let A be the event that the first die is a 3 and B be the event that the sum of the two dice is 8.

Chapter 6

Modeling Random Events:
The Normal and Binomial Models

Table Number:_____   Group Name: _____

Group Members:_____   _____   _____

Discrete Probability Distribution: Ichiro's Hit Parade

In the 2004 baseball season, Ichiro Suzuki of the Seattle Mariners set the record for the most hits in a season with 262 hits. In the following probability distribution, the random variable represents the number of hits Ichiro obtained in a game.

X	P(X)
0	0.1677
1	0.3354
2	0.2857
3	0.1491
4	0.0373
5	0.0248

a. Verify this is a discrete probability distribution

b. Draw a probability histogram, be sure to label your axes.

c. Compute the mean (the expected value) of the random variable and interpret in a complete sentence the mean of the distribution.

d. What is the probability that in a randomly selected game Ichiro got two hits?

e. What is the probability that in a randomly selected game Ichiro got more than one hit? At least three hits?

Table Number:_____ Group Name: _____

Group Members:_____ _____ _____

Expected Value/Mean of a Probability Distribution Practice Set

1. Recall that a roulette wheel has 38 slots. Eighteen are red, 18 are black, and 2 are green. You can bet on 6 different numbers. If any of them comes up, you receive $6 back for each $1 bet. What is the expected loss on a $1 bet? Please round the answer to two decimal places.

    a. $0.01
    b. $0.08
    c. $0.07
    d. $0.05
    e. $0.04

2. In a gambling game, you receive a payoff of $82 if you roll a sum of 4, and $7 if you roll a sum of 7 on two dice. Otherwise, you receive no payoff. What is the average payoff per play?

    a. $8
    b. $9
    c. $6
    d. $12
    e. $5

3. In a carnival game, you roll 2 dice. If the sum is 5, you receive a $6 payoff. If the sum is 10, you receive a $13 payoff.
    What is the expected payoff?

    a. $2.05
    b. $1.35
    c. $1.75
    d. $1.55
    e. $1.85

4. In a gambling game, you pick 1 card from a standard deck. If you pick an ace, you win $10. If you pick a picture card (J, Q, or K), you win $5. Otherwise, you win nothing. How much should a carnival booth charge you to play this game if they want an average profit of $0.60 per game? (hint: first find the average payout)

    a. $2.12
    b. $2.62
    c. $2.72
    d. $2.22
    e. $2.52

Numeric Response

5. You have a job working for a mathematician. She pays you each day according to what card you select from a bag. Two of the cards say $220, five of them say $100, and three of them say $50. What is your expected (average) daily pay?

$ _____ per day

6. An apartment complex has 20 air conditioners. Each summer, a certain number of them have to be replaced.

Number of Air Conditioners Replaced	Probability
0	0.21
1	0.37
2	0.13
3	0.11
4	0.11
5	0.07

What is the expected number of air conditioners that will be replaced in the summer?

7. In a gambling game, you receive a payoff of $46 if you roll a sum of 10, and $7 if you roll a sum of 7 on two dice. Otherwise, you receive no payoff. What is the average payoff per play?

$ _____

Chapter 6   109

Table Number:_____   Group Name: _____

Group Members:_____   _____   _____

What is Normal?

Part I: Making Predictions

Consider the body measurements in the data set BODYMEAS on Stat Crunch

- *Height*
- *Weight*
- *Leg length*
- *Waist circumference*
- *Thigh circumference*

1. Which variables do you expect to have a normal distribution? Why did you pick these?

Part II: Using *StatCrunch* Examine Normal Distributions

Launch *StatCrunch* and access the *BODYMEAS* data set and **generate graphs and summary statistics** for the variables you selected in problem 1.

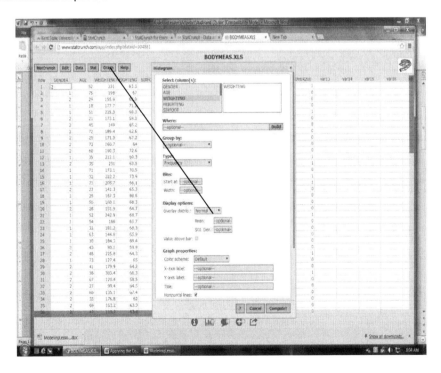

2. Which of those variables appear to be normally distributed? Explain.

3. Pick one distribution that appears to be normally distributed.
   - Draw a picture of the graph for this variable.
   - Label the mean.
   - Mark two standard deviations in each direction from the mean.

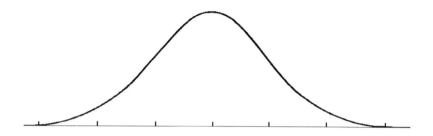

4. What is **your** measurement for this variable? (e.g., what is your own height?)

5. Mark your score on the graph. Are you close to center? In the tails? An outlier?

6. Find the *z*-score for your body measurement for that variable.

7. What does this *z*-score tell you about the location of your body measurement relative to the mean?

Part III: Using a Web Applet to Examine Normal Distributions

- Open StatCrunch
- Click on Stat, Calculators, then Normal. This will open the java applet

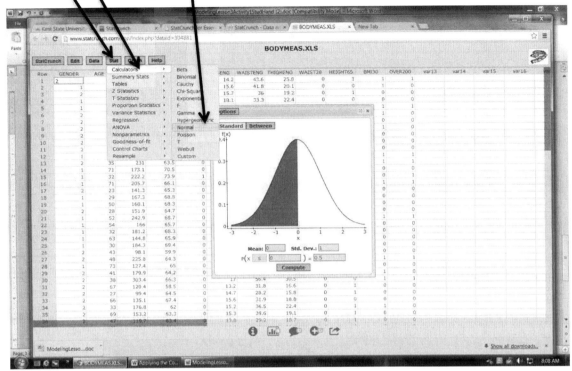

Using the variable you selected in problem 5, enter the mean and standard deviation of that variable found using *StatCrunch* into the proper boxes on the applet.

8. Use the applet to find the proportion of the distribution that is *greater than* your measurement.

9. Does this proportion make sense given the area that is shaded in the applet? Explain.

10. Use the applet to find the proportion of the distribution that is *less than* your measurement.

Reference
Garfield, J., Zieffler, A., & Lane-Getaz, S. (2005). EPSY 3264 Course Packet, University of Minnesota, Minneapolis, MN.

Table Number:_____ Group Name: _____

Group Members:_____ _____ _____

Normal Distribution Applications

A standardized measure of achievement motivation is normally distributed, with a mean of 35 and a standard deviation of 14. Higher scores correspond to more achievement motivation.

1. Draw a picture of this distribution. (Be sure to label the mean and three standard deviations in each direction.)

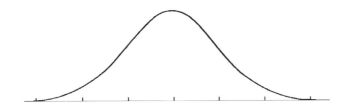

2. Gerry scored 49 on this exam. Mark this score on the distribution you drew in Question 1.

3. Gerry scored higher than what proportion of the population? _____ In your handbook, shade the area that shows this probability and then describe your shading below.

4. 2.5% of the students scored higher than Elaine. What was her achievement motivation score? In your handbook, shade the area 2.5% higher than Elaine and then describe your shading below.

5. The distribution of heights of adult men is approximately normal with a mean of 69 inches and a standard deviation of 2 inches. Bob's height has a z-score of -0.5 when compared to all adult men. Which of the following is true and why?

    A. Bob is shorter than 69 inches tall.
    B. Bob's height is half of a standard deviation below the mean.
    C. Bob is 68 inches tall.
    D. All of the above.

6. Chris is enrolled in a college algebra course and earned a score of 260 on a math achievement test that was given on the first day of class. The instructor looked at two distributions of scores, one for all freshmen who took the test, and the other for students enrolled in the algebra course. Both are approximately normally distributed & have the same mean, but the distribution for the algebra course has a smaller standard deviation. A z-score is calculated for Chris' test score in both distributions (all freshmen & all freshmen taking algebra). Given that Chris' score is well above the mean, which of the following would be true about these two z-scores?

    A. The z-score based on the distribution for the algebra students would be higher.
    B. The z-score based on the distribution for all freshmen would be higher.
    C. The two z-scores would be the same.

7. Explain your answer to Question 6.

8. The average height for all females in the U.S. in *inches* is 65 with a standard deviation of 2.5 inches. Kylee is 68 inches tall, and Michelle is 62 inches tall. Draw a picture of this distribution in your handbook then describe your sketch in the space below. (Be sure to label the mean and three standard deviations in each direction.)

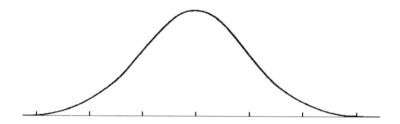

9. What proportion of U.S. females are taller than Kylee?

10. What proportion of U.S. females are shorter than Michelle?

11. What proportion of U.S. females has a height between Michelle and Kylee?

12. SAT I math scores are scaled so that they are approximately normal and the mean is about 511 and the standard deviation is about 112. A college wants to send letters to students scoring in the top 20% on the exam. What SAT I math score should the college use as the dividing line between those who get letters and those who do not?

Reference
Garfield, J., Zieffler, A., & Lane-Getaz, S. (2005). EPSY 3264 Course Packet, University of Minnesota, Minneapolis, MN.

Table Number:_____   Group Name: _____

Group Members:_____   _____   _____

Binomial or Not?

In each of the following cases, state whether or not the process describes a binomial random variable. If it is a binomial, give the value of n and p.

1. Count the number of times a soccer player scores in five penalty kicks against the same goalkeeper. Each shot has a $\frac{1}{3}$ probability of scoring. Circle one of the following, then explain your reasoning.

   YES, BINOMIAL n = _____   p = _____   NO, NOT BINOMIAL
   EXPLAIN:

2. Count the number of times a coin lands heads before it lands tails. Circle one of the following, then explain your reasoning.

   YES, BINOMIAL n = _____   p = _____   NO, NOT BINOMIAL
   EXPLAIN:

3. Draw 10 cards from the top of a deck and record the number of cards that are aces.

   YES, BINOMIAL n = _____   p = _____   NO, NOT BINOMIAL
   EXPLAIN:

4. Conduct a simple random sample of 500 registered voters and record whether each voter is Republican, Democrat, or Independent.

   YES, BINOMIAL n = _____   p = _____   NO, NOT BINOMIAL
   EXPLAIN:

5. Conduct a simple random sample of 500 registered voters and count the number that are Democrats.

   YES, BINOMIAL n = _____   p = _____   NO, NOT BINOMIAL
   EXPLAIN

6. Randomly select one registered voter from each of the 50 US states and count the number that are Democrats.

   YES, BINOMIAL n = _____   p = _____   NO, NOT BINOMIAL
   EXPLAIN:

Table Number:_____   Group Name: _____

Group Members:_____   _____   _____

Binomial Probability Distribution - Practice with Vocabulary

1. According to the Beacon Journal, 40% of bicycles stolen in Akron are recovered. You have a random sample of 6 bicycles. Complete the table for *n* (sample size), *p* (the unchanging probability of success), and *x* (the number of successes). From this information, write how you would enter it into your calculator.

Vocabulary	n (sample size)	P (probability of success)	What values are implied in each of these phrases?	Write the TI*84 calculator command
"exactly 4"				
"less than 4"				
"at least 4"				
"more than 4"				
"at most 4"				

2. Suppose that for a given dog breed, the probability of giving birth to a male puppy is 48%. Suppose also that one particular female dog had a litter of 5 puppies. Complete the table for *n* (sample size), *p* (the unchanging probability of success), and *x* (the number of successes). From this information, write how you would enter it into your calculator.

Vocabulary	n (sample size)	P (probability of success)	What values are implied in each of these phrases?	Write the TI*84 calculator command
"exactly 3"				
"less than 3"				
"at least 3"				
"more than 3"				
"at most 3"				

Table Number:_____   Group Name:_____

Group Members:_____   _____   _____

Binomial Probabilities

1. Suppose that 65% of the families in a town own computers, If ten families are surveyed at random,
   a) What is the probability that at least five own computers? _____
   (SHOW YOUR SET-UP – What is the TI command and arguments)

   b) What is the expected number of families that own computers? _____
   SHOW YOUR SET-UP AND WORK.

2. Ninety percent of a country's population are right-handed.
   a) What is the probability that exactly 29 people in a group of 30 are right-handed? _____
   (SHOW YOUR SET-UP – What is the TI command and arguments)

   b) What is the expected number of right-handed people in a group of 30? _____
   SHOW YOUR SET-UP AND WORK.

3. From the 2010 US Census we learn that 27.5% of US adults have graduated from college. Suppose we take a random sample of 12 US adults.
   a) What is the probability that exactly six of them are college educated? _____
   (SHOW YOUR SET-UP – What is the TI command and arguments)

   b) What is the probability that six or fewer are college educated? _____
   (SHOW YOUR SET-UP – What is the TI command and arguments)

4. In the 2010 census, we learn that 65% of all housing units are owner-occupied while the rest are rented. Suppose we take a random sample of 20 housing units.
   a) What is the probability that exactly 15 of them are owner-occupied? _____
   (SHOW YOUR SET-UP – What is the TI command and arguments)

   b) What is the probability that more than 18 of them are owner-occupied? _____
   (SHOW YOUR SET-UP AND WORK – What is the TI command and arguments)

5. Suppose that a new drug is effective for 65% of the participants in clinical trials. Suppose a group of 15 patients take this drug.
   a) What is the expected number of patients for whom the drug will be effective? _____
      SHOW WORK PLEASE!

   b) What is the probability that the drug will effective for less than half of them? _____

   c) What is the probability that the drug will be effective for more than 75% of them?

6. Suppose that a committee has 10 members. The probability of any member attending a randomly chosen meeting is 0.9. The committee cannot do business if more than 3 members are absent. What is the probability that 7 or more members will be present on a given date?
   SHOW WORK AND YOUR TI COMMAND AND ARGUMENTS. _____

7. During the 2014-2015 NBA (Basketball) season, LeBron James of the Cleveland Cavaliers had a free throw shooting percentage of 0.710. Assume that the probability LeBron makes any given free throw is fixed at 0.710 and that free throws are independent. SHOW ALL WORK FOR EACH QUESTION INCLUDING THE TI_COMMAND AND ARGUMENTS.

   a) If LeBron shoot 8 free throws in a game, what is the probability that he make at least 7 of them? _____

   b) If he shoots 80 free throws in the playoffs, what is the probability that he makes at least 70 of them?
      _____

   c) If he shoots 8 free throws in a game, what is the expected number of free throws that he will make?
      _____

   d) If he shoots 80 free throws in the playoffs, what is the standard deviation for the number of free throws he makes during the playoffs?

      _____

Chapter 7

Survey Sampling and Inference

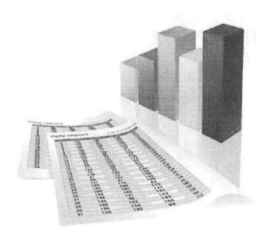

Table Number:_____   Group Name: _____

Group Members:_____   _____   _____

Biased Sampling

Each of the surveys mentioned below has bias. (a) Determine the type of bias and (b) suggest a remedy. Types of bias include *sampling bias (convenience sample, for eg.)*, *voluntary response bias*, *non-response bias*, and *poorly worded question*. Some surveys may include more than one of these.

1. PEARLS BEFORE SWINE

   a) Type:

   b) Remedy:

2. A retail store manager wants to conduct a study regarding the shopping habits of its customers. He selects the first 60 customers who enter his store on a Saturday morning.

   a) Type:

   b) Remedy:

3. A polling organization conducts a study to estimate the percentage of households that speaks a foreign language as the primary language. It mails a questionnaire to 1023 randomly selected households throughout the United States and asks the head of household if a foreign language is the primary language spoken in the home. Of the 1023 households selected, 12 responded.

   a) Type:

   b) Remedy:

4. A newspaper article reported, "The *Cosmopolitan* Magazine survey of more than 5000 Australian women aged 18-34 found about 42% considered themselves overweight or obese."

   a) Type:

   b) Remedy:

5. A researcher asks a random sample of students at the library on a Friday night, "How many hours a week do you study?" to collect data to estimate the average number of hours a week that all college students study.

   a) Type:

   b) Remedy:

6. Take 10 apples off the top of a truckload of apples and measure the amount of bruising on those apples to estimate how much bruising there is, on average, in the whole truckload.

   a) Type:

   b) Remedy:

7. One of the daily polls on *CNN.com* during June 2011 asked, "Does physical beauty matter to you?" Of 38,485 people responding, 79% said yes and 21% said no.

   a) Type:

   b) Remedy:

8. Send an email to a random sample of students at a university asking them to reply to the question: "Do you think this university should fund an ultimate Frisbee team?" A small number of students reply. Use the replies to estimate the proportion of all students at the university who support this use of funds.

   a) Type:

   b) Remedy:

9. A magazine is conducting a study on the effects of infidelity in a marriage. The editors randomly select 400 women whose husbands were unfaithful and ask, "Do you believe a marriage can survive when the husband destroys the trust that must exist between husband and wife?"

   a) Type:

   b) Remedy

Table Number:_____   Group Name: _____

Group Members:_____   _____   _____

Sampling Distribution of Sample Proportions

Purpose: to help you understand what a **sampling distribution** is.

Consider a small population of 5 students. Of these five students, two are female and three are male. Let's refer to them as F1, F2, M1, M2, and M3. Note that the proportion of females in this population is 0.40.

1. **SAMPLE SIZE 2**. Suppose now we want to consider samples of this population with sample size 2, i.e. $n=2$. List ALL the possible samples below. (You should have 10 different samples with $n=2$).

In the table below, list the sample and the proportion of females in the sample;

SAMPLE	PROPORTION OF FEMALES in the sample (*statistic*), $\hat{p}$	PROPORTION OF FEMALES *in the population* (*parameter*), p.

Which can change from sample to sample, the statistic or parameter? _____

Which always stays the same from sample to sample, the statistic or parameter? _____

Write the probability of each $\hat{p}$ in the table below

Value of $\hat{p}$	Probability of $\hat{p}$
0	
0.50	
1.00	

Find the mean, or expected value, of this probability distribution and write it here:_____

This probability distribution of the sample proportions, $\hat{p}$, (a statistic) is called a **sampling distribution.**

Now sketch a probability histogram of $\hat{p}$, the proportion of females you obtained in each sample.

2. **SAMPLE SIZE 3**. Suppose now we randomly select a sample of 3 students. List ALL the possible outcomes below. (You should have 10 different samples with $n = 3$).

In the table below, list the sample and the proportion of females in the sample;

SAMPLE	PROPORTION OF FEMALES in the sample (*statistic*), $\hat{p}$	PROPORTION OF FEMALES in the population (*parameter*), p.

Which can change from sample to sample, the statistic or parameter? _____

Which always stays the same from sample to sample, the statistic or parameter? _____

Write the probability of each $\hat{p}$ in the table below.

Value of $\hat{p}$	Probability of $\hat{p}$
0	
$1/3$	
$2/3$	

Find the mean, or expected value, of this probability distribution and write it here: _____

This probability distribution of the sample proportions (a statistic) is called a **sampling distribution.**

Now sketch a probability histogram of $\hat{p}$, the proportion of females you obtained in each sample.

3. **SAMPLE SIZE 4.** Suppose now we randomly select a sample of 4 students. List ALL the possible outcomes below. (You should have 5 different samples with $n = 4$).

In the table below, list the sample and the proportion of females in the sample;

SAMPLE	PROPORTION OF FEMALES in the sample (**statistic**), $\hat{p}$	PROPORTION OF FEMALES in the population (**parameter**), p.

Which can change from sample to sample, the statistic or parameter? _____

Which always stays the same from sample to sample, the statistic or parameter? _____

Write the probability of each $\hat{p}$ in the table.

Value of $\hat{p}$	Probability of $\hat{p}$
0.25	
0.5	

Now sketch a probability histogram of $\hat{p}$, the proportion of females you obtained in each sample.

Complete: a **sampling distribution** is _____

No matter how many different samples we take, the value of _____ is always the same, but the value of _____ can change from sample to sample.

The expected value, or mean, of the sampling distribution is _____. This means that on average, the estimator, $\hat{p}$, is the same as the population parameter. **Bias** of an estimator is measured as the distance between the mean (expected value) of the sampling distribution and the population parameter. Thus, our estimator has **no bias**. (This is a good thing!)

The **precision** of an estimate is reflected in the spread of the sampling distribution. The standard deviation of the sampling distribution measures this spread and thus is a measure of the estimator's ($\hat{p}$'s) precision. The standard deviation of a sampling distribution is called the **standard error.**

Take another look at the sampling distributions above. Comment about the change in the spread of the distributions as the sample size increases.

Reese's Pieces Part 1

Part 1: Making Conjectures about Samples

Reese's Pieces candies have three colors: Orange, brown, and yellow. *Which color do you think has more candies (occurs more often) in a package*: Orange, brown or yellow?

1. Guess the proportion of each color in a bag:

Color	Orange	Brown	Yellow
Predicted Proportion			

2. If each student in the class takes a sample of 25 Reese's Pieces candies, would you expect every student to have the same number of orange candies in their sample? Explain.

3. Make a conjecture: Pretend that 10 students each took samples of 25 Reese's Pieces candies. Write down the number of orange candies you might expect for these 10 samples:

These numbers show the **variability** you would expect to see in the number of orange candies in 10 **samples** of 25 candies.

You will be given a cup that is a **random sample** of Reese's Pieces candies. Count out 25 candies from this cup without paying attention to color. In fact, try to IGNORE the colors as you do this.

4. Now, count the colors for your sample and fill in the chart below:

	Orange	Yellow	Brown	Total
Number of candies				
Proportion of candies (Divide each number by 25)				

Record both the number and proportion of orange candies on the board.

Repeat the above four more times. For each trial, record the number and proportion of **orange** Reese's Pieces.

	Trial 1	Trial 2	Trial 3	Trial 4	Trial 5
Number of orange candies					
Proportion of orange candies (Divide each number by 25)					

5. Now that you have taken samples of candies and see the proportion of orange candies, make a second conjecture: *If you took a sample of 25 Reese's Pieces candies and found that you had only 5 orange candies, would you be surprised?* Do you think that 5 is an unusual value?

6. Record the proportions of orange candies in your samples on the dotplot below. Then ask one member of your group to create a dotplot with all group member's proportions. That person is to come to the whiteboard and add your group's proportions to the whole class' dotplot.

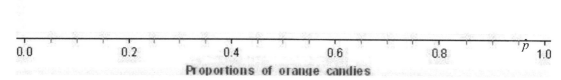

Figure 1: Dot plot for the proportion of orange candies. Your (or your group's) samples.

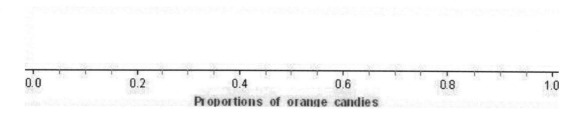

Figure 2: Dot plot for the proportion of orange candies. Samples from entire class

Chapter 7   129

Table Number:_____   Group Name: _____

Group Members:_____   _____   _____

Reese's Pieces Part 2; Simulate the Sampling Process

You will now simulate additional data and tie this activity to the *Simulation Process Model* (SPM).

- Go to our Canvas course homepage, then to the "Modules," then "Chapter 7," then scroll down a bit to the "Reese's Pieces Simulation."
- The first item under the "Reese's Pieces Simulation" heading is the APP for Reese's Pieces Simulation. Click on it.

You will see a big container of colored candies that represents the POPULATION of Reese's Pieces candies.

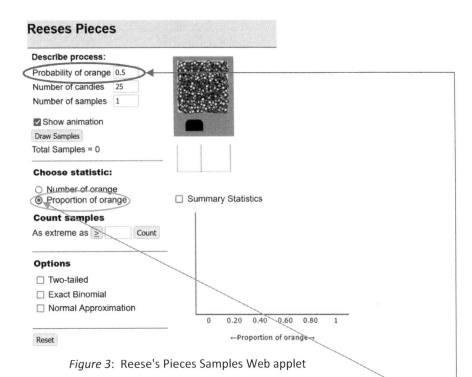

Figure 3: Reese's Pieces Samples Web applet

1. The statistic we will be working with is the **Proportion of orange**, so please click the radio button here so that the app deals with our statistic of interest.

   In addition, note that the probability of getting an orange candy is already given to be 0.5 in the first line. This means that the proportion of orange candies in the population is 0.5. Thus, the app gives us the **parameter**. Normally, we don't know the parameter value but we have to assume a value for the simulation to work.

   Be sure the box next to "Show animation" is checked.

2. How does 0.5 compare to the proportion of orange candies in your sample? Explain.

3. How does it compare to the center of the class's distribution? Does it seem like a plausible value for the population proportion of orange candies? Explain.

Simulation
- Click on the *"Draw Samples"* button in the Reese's Pieces applet. One sample of 25 candies will be taken and the proportion of orange candies for this sample is plotted on the graph.
- Repeat this again. (Draw a second sample.)

4. Do you get the same or different values for each sample proportion?

5. How do these numbers compare to the ones our class obtained?

6. How close is each **sample statistic** (proportion) to the **population parameter**?

Further Simulation
- Uncheck the *"Animate"* box.
- Change the number of samples (*num samples*) to 500.
- Click on the *"Draw Samples"* button, and see the distribution of sample statistics (in this case proportions) build.

7. Describe the shape, center and spread of the distribution of sample statistics. *Important question!*
Shape:

Center:

Spread:

8. How does this distribution compare to the one our class constructed on the board in terms of shape? Center? Spread?

9. Where does the value of 0.2 (i.e., 5 orange candies) fall in the distribution of sample proportions? Is it in the tail or near the middle? Does this seem like a rare or unusual result?

10. a) Are the conditions met in order to conclude that the sampling distribution is approximately Normal? (Use the value for p given in the simulation.)

    Condition 1:

    Condition 2:

    Condition 3:

   b) Find the mean and standard deviation for the sampling distribution:

    mean =

    standard deviation =

   c) Find the probability that the proportion of orange candies in a random sample of 25 Reese's pieces will be less than .2

When we generate sample statistics and graph them, we are generating an estimated **sampling distribution**, or a distribution of the sample statistics. It looks like other distributions we have seen of raw data.

Reference

Rossman, A., & Chance, B. (2002). A data-oriented, active-learning, post-calculus introduction to statistical concepts, applications, and theory. In B. Phillips (Ed.), *Proceedings of the Sixth International Conference on Teaching of Statistics*, Cape Town. Voorburg, The Netherlands: International Statistical Institute. Retrieved September 28, 2007, from http://www.stat.auckland.ac.nz/~iase/publications/1/3i2_ross.pdf

Table Number:_____   Group Name: _____

Group Members:_____   _____   _____

Reese's Pieces Part 3; Examine the Role of Sample Size

Next we consider what will happen to the distribution of sample statistics if we change the number of candies in each sample (change the sample size).

Make a Conjecture

1. What do you think will happen to the distribution of sample proportions if we change the sample size to 50? Explain.

2. What do you think will happen if we change the sample size to 100? Explain.

Test your conjecture. Set the number of samples (*num samples*) in the applet to 500.
  A. Keep the *"sample size"* in the Reese's Pieces applet at 25.
  B. Change the *"sample size"* in the Reese's Pieces applet to 50.
  C. Change the *"sample size"* in the Reese's Pieces applet to 500.
  D. Change the *"sample size"* in the Reese's Pieces applet to 5.

3. As the ***sample size*** increases, what happens to the spread of the distribution?

4. Now, describe the effect of sample size on the distribution of sample statistics in terms of shape, center and spread.

When we generate sample statistics and graph them, we are generating an estimated ***sampling distribution***, or a distribution of the sample statistics. It looks like other distributions we have seen of raw data.

Reference

Rossman, A., & Chance, B. (2002). A data-oriented, active-learning, post-calculus introduction to statistical concepts, applications, and theory. In B. Phillips (Ed.), *Proceedings of the Sixth International Conference on Teaching of Statistics,* Cape Town. Voorburg, The Netherlands: International Statistical Institute. Retrieved September 28, 2007, from http://www.stat.auckland.ac.nz/~iase/publications/1/3i2_ross.pdf

Chapter 7   134

Table Number:_____   Group Name: _____

Group Members:_____   _____   _____

Reese's Pieces Summary Page

Part I

1. Do you know the proportion of orange candies in the population? _____ In the sample? _____
   Which one can we always calculate?_____ Which one do we have to estimate? _____

2. Describe the variability of the distribution of sample proportions of the whole class (the one on the whiteboard) in terms of shape, center, and spread. *This is a most important question!*
   Shape:

   Center:

   Spread:

3. Where does the value of 0.2 (i.e. 5 orange candies) fall in the distribution of sample proportions? Is it in the tail or near the middle? Does this seem like a rare or unusual result?

4. Find the probability that the proportion of orange candies in a random sample of 25 Reese's pieces will be less than .2

Part II & III.

5. What happens to the mean, standard deviation, and distribution graph as you increase the number of samples?

6. As the sample SIZE increases, what happens to the distance the sample statistics are to the population parameter?

7. Describe the effect of sample size on the distribution of sample statistics in terms of
   a) Shape

   b) Center

   c) Spread.

Table Number:_____   Group Name: _____

Group Members:_____   _____   _____

Standard Error Extra Practice

Find the standard error for each of the following scenario. Show your work (if possible) and place your answer on the line to the right. Round your answer to 4 decimal places.

1. $p=.19$ and $n=10$. Population size is 100   1._____

2. $p=.19$ and $n=50$. Population size is 50000   2._____

3. $p=.19$ and $n=60$. Population size is 600   3._____

4. $p=.19$ and $n=100$. Population size is 1200   4._____

5. $p=.19$ and $n=1000$. Population size is 50,000   5._____

6. $p=.40$ and $n=1000$. Population size is 50,000   6._____

7. $p=.10$ and $n=1000$. Population size is 50,000   7._____

Now answer these: **True or False.**

8. The population size affects the standard error.   8._____
9. The sample size affects the standard error.   9._____
10. The population proportion, p, affects the standard error.   10._____
11. The complement of the population proportion affects the standard error. 11._____
12. The larger the sample, the larger the standard error.   12._____
13. The smaller the sample, the larger the standard error.   13._____
14. The larger the sample, the smaller the standard error.   14._____
15. We use a population proportion to make inferences about a sample.   15._____
16. We use a sample proportion to make inferences about a population.   16._____
17. Explain in your own words how the sample size affects the standard error and why.

Chapter 7

Understanding and Interpreting Confidence Intervals
from Section 3.2 in Lock, Robin et al. Statistics, Unlocking the Power of Data. Wiley Pub. 2013

1. Is a Television Set Necessary?
The percent of Americans saying that a television set is a necessity has dropped dramatically in recent years. In a nationwide survey conducted in May 2010 of 1484 people ages 18 and older living in the continental United States, only 42% say that a television set is a necessity rather than a luxury. The article goes on to say "the margin of error is plus or minus 3.0 percentage points." Use the information from this article to find an interval estimate for the proportion of people 18 and older living in the continental United States who believe that a television set is a necessity.

Interval Estimate Based on a Margin of Error

An **interval estimate** gives a range of plausible values for a population parameter. One common form of an interval estimate is

$$\text{Point estimate} \pm \text{margin of error}$$

where the **margin of error** is a number that reflects the **precision** of the sample statistic as a point estimate for this parameter

---

2. Suppose the results of an election poll show the proportion supporting a particular candidate to be $\hat{p} = 0.54$.

   We would like to know how close the true p is to $\hat{p}$. Two possible margins of error are shown below. In each case, indicate whether we can be reasonably sure that this candidate will win the majority of votes and win the election.

   a. Margin of error is 0.02

   b. Margin of error is 0.10

Confidence Intervals

This "range of plausible values" interpretation for interval estimate can be refined within the notion of *confidence interval*.

Confidence Interval

A **confidence interval** for parameter in an interval computed from sample data by a method that will capture the parameter for a specified proportion of all samples.

The success rate (proportion of all samples whose intervals contain the parameter) is known as the **confidence level.**

95% Confidence Interval Using the Standard Error

If we can estimate the standard error *SE* and if the sampling distribution is relatively symmetric and bell-shaped, a 95% confidence interval can be estimated using

$$\text{Statistic} \pm 2 \cdot SE$$

---

3. Each of the three values listed below is one of the samples shown in the dotplot in the figure. Find a 95% confidence interval using the sample proportion and the fact that the standard error is about 0.03. In each case, also locate the sample proportion on the sampling distribution and indicate whether the 95% confidence interval captures the true population proportion.

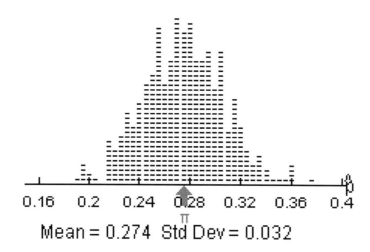

Mean = 0.274 Std Dev = 0.032

a) $\hat{p} = 0.26$

b) $\hat{p} = 0.32$

c) $\hat{p} = 0.20$

4. Construct an interval estimate for p using $\hat{p}=0.37$ with margin of error 0.02.

5. Assume a sampling distribution is symmetric and bell-shaped use the following information to construct a 95% confidence interval if $\hat{p}=0.32$ and the standard error is 0.04.

6. A 95% confidence interval for a proportion is 0.72 to 0.79. Is the value given a plausible value of p?

   a) $p=0.85$

   b) $p=0.75$

   c) $p=0.07$

7. The CDC classifies an adult as "obese" if the Body Mass Index (BMI) ≥ 30. Based on the data from a 2010 survey, a 95% confidence interval for the proportion of all adults living in the US that were obese in 2010, p_{2010} is (0.276, 0.278). Based on the data from a 2000 survey, a 95% confidence interval for the proportion of all adults living in the US that were obese in 2000, p_{2000} is (0.195, 0.199).

   a) Interpret both confidence intervals

   b) Do you think the sample size for the 2000 survey was smaller or larger than the sample size for the 2010 survey? Why?

8. A random sample of $n=1483$ adults in the US were asked whether they consider a car a necessity or a luxury and we find that a 95% confidence interval for the proportion saying that it is a necessity is 0.83 to 0.89. Explain the meaning of this confidence interval in the appropriate context.

Table Number:_____   Group Name: _____

Group Members:_____   _____   _____

Confidence Intervals for a Proportion

Open this applet, which allows you to visually investigate confidence intervals for a proportion. This link is also available in the Chapter 7 Module on Canvas, immediately following GW43.

Simulating Confidence Intervals

Describe process
- Statistic: Proportions
- Distribution: Binomial
- Method: Wald
- π: 0.5
- Sample size (n): 25
- Number of intervals: 50
- [Sample]

Confidence level: 95 % [Recalculate]

Results

[Sort]
[Reset]

Specify the sample size **n** and the population proportion π. (In real life, we usually don't know the population proportion.) To obtain 50 confidence intervals from 50 randomly selected samples, type in "50" in the box to the right of "Intervals." When you click the **Sample** button, 50 separate 95% confidence intervals from 50 randomly selected samples of size **n** will be displayed. Recall that we can use these confidence intervals to approximate the population proportion. Each of these intervals is computed using the standard normal approximation. If an interval does not contain the true proportion, it is displayed in red. Pressing the **Sample** button again will show confidence intervals for a *different set of 50 samples*. Press the **Reset** button to clear existing results and start a new simulation. Things to try with the applet:

1. Simulate 50 intervals with **n** = 25 and **p** = 0.5. How many confidence intervals contain 0.5? What proportion of the 50 confidence intervals is this?

2. Click **Reset** and repeat #1 by clicking **Sample** again. Did you get the same number (and same proportion) of intervals containing **p** as you did in #1? How can this be?

3. Repeat #2 several more times and list the proportion of intervals containing **p** in each one of your simulations in the space below. Please complete at least 5 more simulations.

4. Now let's vary the sample size. Simulate 50 intervals with **n** = 10 and **p** = 0.5.
    a. What proportion of the 95% confidence intervals contain 0.5?

    b. How does the width of these intervals compare to those in # 1 above?

5. Now repeat #4, except with sample size **n** = 100.
    a. What proportion of the 95% confidence intervals contain 0.5?

    b. What do you notice about the width of the confidence intervals compared to those in #1 and #4?

    c. Think about **why** the width of the confidence intervals change with different sample sizes and write your ideas in the space below.

6. Continue experimenting with different sample sizes and different values for **p**.

    a. Comment below about the varying widths of the intervals and any patterns you have noticed relating sample size to the width of the intervals.

    b. Does the number of intervals you create have any effect on the patterns you noticed in part a?

Chapter 8

Hypothesis Testing for Population Proportions

Table Number:_____   Group Name: _____

Group Members:_____   _____   _____

Balancing Coins Part 1

Part 1: Testing Statistical Hypotheses

After talking to several expert coin balancers, we are under the impression that balancing a coin and letting it fall is an "unfair" process. In other words, this does not produce results that are 50% heads and 50% tails. *We will test the hypothesis that the proportion of heads when balancing a coin repeatedly on its edge is not 0.5.*

We will design an experiment in order to be able to make a decision about the two hypotheses. In order to make a decision about the research question, we need four things:

1. **A hypothesis to test**
2. **A sample of data which gives us a sample statistic**
3. **A sampling distribution** for the sample statistic that we obtained. This is the model we would expect if the null-hypothesis is true. We can use the sampling distribution to see how unusual or surprising our sample result is. If our sample result is in one of the tails of the distribution, it would lead us to suspect that our result is surprising given that particular model. This would therefore be evidence against the null-hypothesis.
4. **A decision rule**: How far in the tails does our sample result have to be? The decision rule tells us how far in one of the tails our sample result needs to be for us to decide it is so unusual and surprising that we reject the idea stated in the null-hypothesis.

Writing statistical hypotheses

We basically want to know if we can expect an average of 50% heads if we repeatedly balance a coin ten times, each time counting the number of heads that show or not. We can write those two ideas as follows:

Idea 1:   Balancing a coin is a "fair" process.
Idea 2:   Balancing a coin is an "unfair" process.

We can also write these ideas using more mathematical ideas, as shown below:

H_o : The proportion of heads when we balance a coin repeatedly is 0.5.

H_A : The proportion of heads when we balance a coin repeatedly is not 0.5.

(In other words the proportion is more, or less, than 0.5.)

We call these ideas **'statistical hypotheses'**. **The first idea** states that the coin is just a likely to land heads as it is to land tails, or that there will be an equal number of heads and tails when we balance a coin. This statement is called the **'null hypothesis'** because it represents an *idea of no difference* from the norm or prior belief or *no effect* (e.g., getting the same results as tossing fair coins). The null-hypothesis is labeled by the symbol 'H_0'.

The second idea states that there will **not** be an equal number of heads and tails, something different than the first idea, so it is called the **'alternative hypothesis'**. The symbol used for the alternative hypothesis is 'H_a'. Note that this was our original hypothesis about the proportion of heads.

> We gather evidence (data) to see if we can disprove the null hypothesis. If we do, then we will accept that the alternative hypothesis is true. The decision between the two hypotheses is usually expressed in terms of H_0 (idea # 1). If our data lead us to believe the second idea is true, then we usually say that **'we reject H_0'**.

Gathering evidence (data) to make the decision whether or not to reject H_0

Balance a penny coin. Make sure the head side of the coin is facing you. When it is standing upright on its edge, hit the under-side of the table (right underneath the balanced coin). Record if you got a head or tail. Repeat this process ten times.

Trial	Result (H or T)
1	
2	
3	
4	
5	
6	
7	
8	
9	
10	

How many Heads did you get? _____

Your **sample statistic** is the *proportion of heads* (divide the number of heads by 10).

Write that here: _____

Find an appropriate sampling distribution

What shall we use? Remember that the sampling distribution is consistent with the idea expressed in the null-hypothesis. Since the null-hypothesis is that the proportion of heads is 0.5, we can refer to the sampling distribution we created earlier.

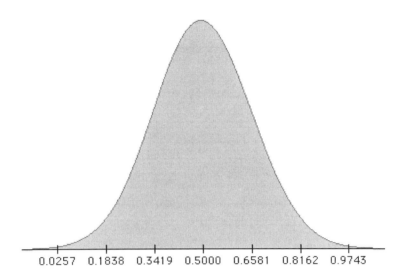

This sampling distribution allows you to compare your sample proportion of heads, based on ten trials, to other sample proportions of heads based on ten trials that you could have obtained, given that the null-hypothesis is indeed true (that balancing a coin repeatedly on its edge produces 50% heads).

Use your sketch to determine whether or not your result is in a tail. Mark your result on your graph. Does your result seem surprising? Explain.

Decision Rule

How unlikely is your sample result? To determine this, do the following:

- Open the *Stat Crunch in MyMathLab*.
- Click on STAT, then PROPORTION STATISTICS, then ONE SAMPLE, then, WITH SUMMARY.
- Enter the number of success and the number of observations.
- Enter in 0.5 after H_o. Use $\neq$ for the alternative hypothesis, H_a. This is an example of a two tail test.
- Click on Compute make note of the results, particularly the **p-value**

The **p-value** is an index of the likelihood of your sample result or a more extreme sample result under the model of the null-hypothesis.

Write your **p-value** here: _____

> This probability of getting the result we got or a more extreme one is called the **p-value.** We can use the p-value to help make a decision about which of our hypotheses seems more likely. *A typical decision rule for p-values is to see if they are smaller than 0.05.* If your p-value is less than 0.05, then it would be evidence against the null hypothesis - you would *reject the null hypothesis*. If your p-value is greater than or equal to 0.05, then it would be evidence for the null hypothesis - you would *fail to reject the null hypothesis*.

How does your p-value compare to 0.05? Is it evidence *for* or *against* the null hypothesis?

What does this suggest about the "fairness" of the process of balancing a coin?

Did all your classmates come to the same conclusion? If not, why did we get different decisions in the same experiment?

> *What does the p-value tell us?* To summarize our results we would say that the probability of getting a sample proportion such as the one we got or a more extreme sample proportion, when the null hypothesis that heads and tails are equally likely is really true is _____. *(Fill in the blank with the p-value.)*

References

Scheaffer, R.L., Watkins, A. Witmer, J., & Gnanadesikan, M., (2004a). *Activity-based statistics: Instructor resources* (2nd edition, Revised by Tim Erickson). Key College Publishing.

Scheaffer, R.L., Watkins, A., Witmer, J., & Gnanadesikan, M., (2004b). *Activity-based statistics: Student guide* (2nd edition, Revised by Tim Erickson). Key College Publishing.

Practice with the null and alternative hypotheses

REMEMBER: Hypothesis testing is a procedure, based on sample evidence and probability, used to test statements regarding a characteristic of one or more populations.

***

Complete:

The _____ is a statement of NO CHANGE, NO EFFECT, NO DIFFERENCE.

The _____ is a statement we are trying to find evidence to support.

***

In problems #1 – 6, the null and alternative hypotheses are given. Determine whether the hypothesis test is left-tailed, right-tailed, or two-tailed. What parameter is being tested?

1. $H_0: \mu = 5$
   $H_a: \mu > 5$

2. $H_0: p = 0.2$
   $H_a: p < 0.2$

3. $H_0: \sigma = 4.2$
   $H_a: \sigma \neq 4.2$

4. $H_0: p = 0.76$
   $H_a: p > 0.76$

5. $H_0: \mu = 120$
   $H_a: \mu < 120$

6. $H_0: \sigma = 7.8$
   $H_a: \sigma \neq 7.8$

In #7 – 13, determine the null and alternative hypotheses.

7. **Teenage Mothers.** According to the U.S. Census Bureau, 10.5% of registered births in the United States in 2007 were to teenage mothers. A sociologist believes that this percentage has increased since then.
   H_0:

   H_a:

8. **Charitable Contributions.** According to the Center on Philanthropy at Indiana University, the mean charitable contribution per household among households with income of $1 million or more in the United States in 2005 was $17,072. A researcher believes that the level of giving has changed since then.

   H_0:

   H_a:

9. **Single-Family Home Price.** According to the National Association of Home Builders, the mean price of an existing single-family home in 2009 was $218,600. A real estate broker believes that because of the recent credit crunch, the mean price has decreased since then.

   H_0:

   H_a:

10. **Fair Packaging and Labeling.** Federal law requires that a jar of peanut butter that is labeled as containing 32 ounces must contain at least 32 ounces. A consumer advocate feels that a certain peanut butter manufacturer is shorting customers by underfilling the jars.

    H_0:

    H_a:

11. **Valve Pressure.** The standard deviation in the pressure required to open a certain valve is known to be $\sigma = 0.7$ psi. Due to changes in the manufacturing process, the quality-control manager feels that the pressure variability has been reduced.

    H_0:

    H_a:

12. **Overweight.** According to the Centers for Disease Control and Prevention, 19.6% of children aged 6 to 11 years are overweight. A school nurse thinks that the percentage of 6- to 11-year-olds who are overweight is higher in her school district.

    H_0:

    H_a:

13. **Cell Phone Service.** According to the *CTIA- The Wireless Association,* the mean monthly cell phone bill was $47.47 in 2010. A researcher suspects that the mean monthly cell phone bill is different today.

    H_0:

    H_a:

For Problems #7 – 13 above state the conclusion based on the results of the test.

14. For the hypotheses in problem 7, the null hypothesis is rejected.

15. For the hypotheses in problem 8, the null hypothesis is not rejected.

16. For the hypotheses in problem 9, the null hypothesis is not rejected.

17. For the hypotheses in problem 10, the null hypothesis is rejected.

18. For the hypotheses in problem 11, the null hypothesis is not rejected.

19. For the hypotheses in problem 12, the null hypothesis is not rejected.

20. For the hypotheses in problem 13, the null hypothesis is rejected.

Table Number:_____   Group Name: _____

Group Members:_____   _____   _____

Types of Errors

In hypothesis testing we need to decide that either H_0 or H_a is true. Obviously we would like to make the correct decision, but we can sometimes make the wrong decision. How would you describe each of the following errors in words in terms of the following situation?

H_0: The defendant is innocent; H_a: The defendant is guilty (not innocent).

1) We decide that H_a is true, but this is the wrong decision because H_0 is true.

2) We decided that H_0 is true, but it is the wrong decision because H_0 is not true.

We call these situations a **type I error** and **type II error** respectively. Of course we would like to keep the chances of making a mistake very small. We usually express our decision in terms of the null hypothesis, H_0. (We either reject or fail to reject H_0.) In the same way, we usually focus on the probability of making type I error (rejecting the null hypothesis when it is true). This is because the null hypothesis reflects a 'status quo' or neutrality situation, and if we reject it we are making a statement saying that something is better or preferred, or worse, or different, depending on the situation.

Consider the following two hypotheses that could be used to examine the quality control process in a parachute factory.

H_0: The parachutes being produced will not open.
H_A: The parachutes being produced will open.

Describe what a type I error would be given these hypotheses. What are the practical implications of making a type I error?

Describe what a type II error would be given these hypotheses. What are the practical implications of making a type II error?

In many situations, making either error doesn't have much practical significance. In the two situations above, though, making one type of error is much more costly than making the other.

Here's another example: When the effectiveness of one medication is studied, (H_0: Medicine A is no more effective than current medications), a ' type I error' would mean that the new medicine is concluded to be better when it actually is no more effective. Type I error is usually considered a serious error and we like to have some control over it.

Reference

Seier, E., & Robe, C. (2002). Ducks and green – an introduction to the ideas of hypothesis testing. *Teaching Statistics, 24*(3), 82-86.

Table Number:_____ Group Name: _____

Group Members:_____ _____ _____

Practice Type I and Type II Errors

REMEMBER: A Type I error is REJECTING A TRUE NULL HYPOTHESIS

The LEVEL OF SIGNIFICANCE is the probability of making a Type I error

A Type II error is NOT REJECTING A FALSE NULL HYPOTHESIS

In each scenario, write a sentence that would illustrate the indicated error.

1. **Teenage Mothers.** According to the U.S. Census Bureau, 10.5% of registered births in the United States in 2007 were to teenage mothers. A sociologist believes that this percentage has increased since then.

    Type I error:

    Type II error:

2. **Charitable Contributions.** According to the Center on Philanthropy at Indiana University, the mean charitable contribution per household among households with income of $1 million or more in the United States in 2005 was $17,072. A researcher believes that the level of giving has changed since then.

    Type I error:

    Type II error:

3. **Single-Family Home Price.** According to the National Association of Home Builders, the mean price of an existing single-family home in 2009 was $218,600. A real estate broker believes that because of the recent credit crunch, the mean price has decreased since then.

    Type I error:

    Type II error:

4. **Fair Packaging and Labeling.** Federal law requires that a jar of peanut butter that is labeled as containing 32 ounces must contain at least 32 ounces. A consumer advocate feels that a certain peanut butter manufacturer is shorting customers by under filling the jars.
   Type I Error:

   Type II Error:

5. **Valve Pressure.** The standard deviation in the pressure required to open a certain valve is known to be $\sigma = 0.7$ psi. Due to changes in the manufacturing process, the quality-control manager feels that the pressure variability has been reduced.
   Type I Error:

   Type II Error:

6. **Overweight.** According to the Centers for Disease Control and Prevention, 19.6% of children aged 6 to 11 years are overweight. A school nurse thinks that the percentage of 6- to 11-year-olds who are overweight is higher in her school district.
   Type I Error:

   Type II Error:

7. **Cell Phone Service.** According to the *CTIA- The Wireless Association,* the mean monthly cell phone bill was $47.47 in 2010. A researcher suspects that the mean monthly cell phone bill is different today.
   Type I Error:

   Type II Error:

Table Number:_____   Group Name: _____

Group Members:_____   _____   _____

Hypothesis Testing for Proportions

1. A 2003 study of dreaming found that out of a random sample of 113 people, 92 reported dreaming in color. However the rate of reported dreaming in color that was established in the 1940s was 0.29. Check to see whether the conditions for using a one-proportion z-test are met assuming the researcher wanted to test to see if the proportion dreaming in color had changed since the 1940's.

2. Suppose you are testing the claim that a coin comes us tails more than 50% of the time when the coin is spun on a hard surface. Steps 1 and 2 of the hypothesis test are given. Suppose that you did this experiment and got 22 tails in 30 spins. Find the value of the test statistic z and the corresponding p-value.

   p is the proportion of tails.

   STEP 1: $H_0: p = 0.50$
   $H_a: p > 0.50$

   STEP 2: Assume that the outcomes are random and the sample size is large enough because both np and $n(1-p)$ are both 15.

   z- statistic (SHOW WORK!!) _____

   p-value: _____ (Use graphing calculator)

SKILL AND DRILL

In problems #3 - #8, test the hypotheses. Find the p-value and indicate whether the researcher should reject or not reject the null hypothesis. Please note that not all significance levels are 0.05.

3. $H_0: p = 0.3$
   $H_a: p > 0.3$

   $n = 200; x = 75; \alpha = 0.05$
   p-value:_____   Circle one: REJECT H_0 DO NOT REJECT H_0

4. $H_0: p = 0.6$
   $H_a: p < 0.6$

   $n = 250; x = 124; \alpha = 0.01$
   p-value:_____   Circle one: REJECT H_0 DO NOT REJECT H_0

5. $H_0: p = 0.55$
   $H_a: p < 0.55$

   $n = 150; x = 78; \alpha = 0.1$
   p-value:_____   Circle one: REJECT H_0 DO NOT REJECT H_0

6. $H_0: p = 0.25$
   $H_a: p < 0.25$

   $n = 400; x = 96; \alpha = 0.1$
   p-value:_____   Circle one: REJECT H_0 DO NOT REJECT H_0

7. $H_0: p = 0.9$
   $H_a: p \neq 0.9$

   $n = 500; x = 440; \alpha = 0.05$
   p-value:_____   Circle one: REJECT H_0 DO NOT REJECT H_0

8. $H_0: p = 0.4$
   $H_a: p \neq 0.4$

   $n = 1000; x = 420; \alpha = 0.01$
   p-value:_____   Circle one: REJECT H_0 DO NOT REJECT H_0

9. In a 2010 poll of 1000 adults, 520 of those polled said that schools should ban all junk food from vending machines in schools. Do a majority of adults (more than 50%) support a ban on junk food? Perform a hypothesis test using a significance level of 0.05. (See p. 342 in your text for all the steps.)

   STEP 1
   H_0:
   H_a:

   STEP 2

   STEP 3

STEP 4: (Circle one)　　REJECT H₀　　　　DO NOT REJECT H₀

Choose the best interpretation of the results you obtained in STEP 4:

I. The percentage of all adults who favor banning is significantly more than 50%

II. The percentage of all adults who favor banning is not significantly more than 50%

Chapter 9

Inferring Population Means

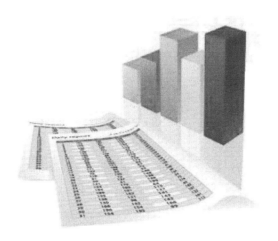

Table Number:_____   Group Name: _____

Group Members:_____   _____   _____

Sampling Distributions of Sample Means

Consider the following quiz scores (out of 10 possible points) for 5 students in our class: 7, 8, 6, 10, 4. This set of five quiz scores is our population.

1. Find the following parameters of this population:
    a. Mean, $\mu =$ _____
    b. Standard Deviation, $\sigma =$ _____

2. **SAMPLING DISTRIBUTION FOR SIZE N = 3.** Suppose now we want to consider all samples of this population with sample size 3, i.e. $n = 3$. There are 10 different such samples. I've listed them in the table below.

    a. Find the mean, $\bar{x}$, of each sample.

SAMPLE	MEAN of the sample (*statistic*), $\bar{x}$	|Distance| of each $\bar{x}$ from μ $\|\bar{x} - \mu\|$
7,8,6		
7,8,10		
7,8,4		
7,6,10		
7,6,4		
7,10,4		
8,6,10		
8,6,4		
8,10,4		
6,10,4		

> This distribution of all possible sample means (a statistic) is called a **sampling distribution.**

   b. Find the mean of this distribution of sample means and write it here: _____ We use the symbol, $\mu_{\bar{x}}$ to represent this "mean of sample means." What do you notice about $\mu_{\bar{x}}$ and the population mean, μ ?

   c. Find the standard deviation of the distribution of $\bar{x}$ and write it here:_____. What do you notice about $\sigma_{\bar{x}}$ and σ ? _____

3. **SAMPLING DISTRIBUTION FOR SIZE N = 4.** Suppose now we want to consider all samples of this population with sample size 4, i.e. $n=4$. There are 5 different such samples. I've listed them in the table below.

   a. Find the mean, $\bar{x}$, of each sample.

SAMPLE	MEAN of the sample (*statistic*), $\bar{x}$	|Distance| of each $\bar{x}$ from μ $\|\bar{x} - \mu\|$
7,8,6,10		
7,8,6,4		
7,8,10,4		
7,6,10,4		
8,6,10,4		

   b. Find the mean of this distribution of sample means and write it here: _____ We use the symbol, $\mu_{\bar{x}}$ to represent this "mean of sample means." What do you notice about $\mu_{\bar{x}}$ and the population mean, μ ?

   c. Find the standard deviation of the distribution of $\bar{x}$ and write it here: _____ . What do you notice about $\sigma_{\bar{x}}$ and σ ? _____

4. **SAMPLING DISTRIBUTION FOR SIZE N = 2.** Suppose now we want to consider all samples of this population with sample size 2, i.e. $n=2$. There are 10 different such samples. List them in the table below.

   a. Find the mean, $\bar{x}$, of each sample.

SAMPLE	MEAN of the sample (*statistic*), $\bar{x}$	|Distance| of each $\bar{x}$ from μ $\|\bar{x} - \mu\|$

   b. Find the mean of this distribution of sample means and write it here: _____ We use the symbol, $\mu_{\bar{x}}$ to represent this "mean of sample means." What do you notice about $\mu_{\bar{x}}$ and the population mean, μ ?

   c. Find the standard deviation of the distribution of $\bar{x}$ and write it here: _____ . What do you notice about $\sigma_{\bar{x}}$ and σ ? _____

CONCLUSION: Complete the following with either $=, \neq, <, >$: $\mu_{\bar{x}}$ _____ μ ; $\sigma_{\bar{x}}$ _____ σ ;

Table Number:_____   Group Name: _____

Group Members:_____   _____   _____

Estimating Word Lengths

To understand more about confidence intervals, we are going to return to the *Gettysburg Address* activity, in which we sampled words from the Gettysburg Address. We will use the Gettysburg Address as the population, and take random samples and construct confidence intervals so that we can see how they behave and how to interpret them.

Research Question: What is a good estimate for the *average* word length for *all of the words* in the Gettysburg Address?

Use the Gettysburg Address applet (located on the course website in the Chapter 9 folder or <u>here</u>) to take a random sample of 25 words. Set the sample size to 25 and number of samples to 1. This will draw a random sample of 25 words from the Gettysburg Address and give you the sample mean. Write the sample mean here:

Mean: _____ and the sample standard deviation: _____

Now, find a 95% confidence interval to estimate the true mean word length for all of the words in the Gettysburg Address.

Write the correct *t* multiplier here (Use the *t*-Distribution Critical Values table on p. A-7 in your text.) _____

Margin of error: (Show work please!) _____

Interval: (Show work please!) _____

1. Provide an interpretation of the results. Remember that you will need to report the *interval estimate*, and the *level of confidence* in your interpretation.

2. Did the interval you found include the true mean word length of 4.29?

3. What percentage of all the intervals in the class would you expect to NOT overlap the population mean? Explain.

Repeat the above to find another 95% confidence interval to estimate the true mean word length for all of the words in the Gettysburg Address.

Use the Gettysburg Address applet (located on the course website in the Chapter 9 folder or <u>here</u>) to take another random sample of 25 words. Set the sample size to 25 and number of samples to 1.

Mean: _____ and the sample standard deviation: _____

Now, find the 95% confidence interval.

Write the correct *t* multiplier here (Use the *t*-Distribution Critical Values table on p. A-7 in your text.) _____

Margin of error: (Show work please!) _____

Interval: (Show work please!) _____

4. Provide an interpretation of the results. Remember that you will need to report the *interval estimate*, and the *level of confidence* in your interpretation.

5. Did the interval you found include the true mean word length of 4.29?

6. What percentage of all the intervals in the class would you expect to NOT overlap the population mean? Explain.

Repeat the above, except now find a 90% confidence interval to estimate the true mean word length for all of the words in the Gettysburg Address.

Use the Gettysburg Address applet (located on the course website in the Chapter 9 folder or here) to take another random sample of 25 words. Set the sample size to 25 and number of samples to 1.

Mean: _____ and the sample standard deviation: _____

Now, find the 90% confidence interval.

Write the correct *t* multiplier here (Use the *t*-Distribution Critical Values table on p. A-7 in your text.) _____

Margin of error: (Show work please!) _____

Interval: (Show work please!) _____

7. Provide an interpretation of the results. Remember that you will need to report the *interval estimate*, and the *level of confidence* in your interpretation.

8. Did the interval you found include the true mean word length of 4.29?

9. What percentage of all the intervals in the class would you expect to NOT overlap the population mean? Explain.

Reference

Garfield, J., & Zieffler, A. (2007). EPSY 3264 Course Packet, University of Minnesota, Minneapolis, MN.

Table Number:_____ Group Name: _____

Group Members:_____ _____ _____

Confidence Intervals for a Mean

Open this applet, which allows you to visually investigate confidence intervals for a proportion. This link is also available in the Chapter 9 Module on Canvas, immediately following GW53. Choose items in the drop-down boxes on the app so that your screen looks like this:

Simulating Confidence Intervals

Describe process
- Statistic: Means
- Distribution: Normal
- Method: z with sigma
- Population mean (μ): 75
- Population SD (σ): 4
- Sample size (n): 25
- Number of intervals: 50
- Sample

Confidence level: 95 % Recalculate

Results

Sort
Reset

Specify the sample size **n,** the true mean μ, and the population standard deviation. (In real life, we usually don't know these parameters.) Then specific the number of intervals you want to view (25 is a good number with which to start). To obtain 50 confidence intervals from 50 randomly selected samples, type in "50" in the box to the right of "Intervals." When you click the **Sample** button, 50 separate 95% confidence intervals from 50 randomly selected samples of size **n** will be displayed. Recall that we can use these confidence intervals to approximate the population mean, μ. Each of these intervals is computed using the standard normal approximation. If an interval does not contain the true population mean, it is displayed in red. Pressing the **Sample** button again will show confidence intervals for a *different set of 50 samples.* Press the **Reset** button to clear existing results and start a new simulation. Things to try with the applet:

1. Simulate 50 intervals with **n** = 25 and $\mu = 75$. How many confidence intervals contain 75? What proportion of the 50 confidence intervals is this?

2. Click **Reset** and repeat #1 by clicking **Sample** again. Did you get the same number (and same proportion) of intervals containing 75 as you did in #1? How can this be?

3. Repeat #2 several more times and list the proportion of intervals containing 75 in each one of your simulations in the space below. Please complete at least 5 more simulations.

4. Now let's vary the sample size. Simulate 50 intervals with **n** = 10 and $\mu = 75$.
    a. What proportion of the 95% confidence intervals contain 75?

    b. How does the width of these intervals compare to those in # 1 above?

5. Now repeat #4, except with sample size **n** = 100.
    a. What proportion of the 95% confidence intervals contain 75?

    b. What do you notice about the width of the confidence intervals compared to those in #1 and #4?

    c. Think about **why** the width of the confidence intervals change with different sample sizes and write your ideas in the space below.

6. Continue experimenting with different sample sizes and different values for μ.

    a. Comment below about the varying widths of the intervals and any patterns you have noticed relating sample size to the width of the intervals.

    b. Does the number of intervals you create have any effect on the patterns you noticed in part a?

Table Number:_____   Group Name: _____

Group Members:_____   _____   _____

SKILL AND DRILL CONFIDENCE INTERVALS FOR A MEAN.

Construct a 95% confidence interval given each of the following sample means and sample sizes. Recall that the format of a confidence interval for a mean is: $\bar{x} \pm (t^*) SE_{Est}$ where $SE_{Est} = \dfrac{s}{\sqrt{n}}$ and $s =$ the sample standard deviation. Assume conditions of the CLT are satisfied.

1. $\bar{x} = 90$; $s = 2.5$; $n = 10$ $t^* =$ _____ SE_{Est} _____ Interval:_____

2. $\bar{x} = 90$; $s = 2.5$; $n = 25$ $t^* =$ _____ SE_{Est} _____ Interval:_____

3. $\bar{x} = 90$; $s = 2.5$; $n = 100$ $t^* =$ _____ SE_{Est} _____ Interval:_____

4. $\bar{x} = 90$; $s = 3.5$; $n = 100$ $t^* =$ _____ SE_{Est} _____ Interval:_____

5. $\bar{x} = 4.5$; $s = 0.75$; $n = 25$ $t^* =$ _____ SE_{Est} _____ Interval:_____

6. $\bar{x} = 4.5$; $s = 0.75$; $n = 100$ $t^* =$ _____ SE_{Est} _____ Interval:_____

7. $\bar{x} = 4.5$; $s = 0.75$; $n = 500$ $t^* =$ _____ SE_{Est} _____ Interval:_____

8. $\bar{x} = 4.5$; $s = 0.25$; $n = 500$ $t^* =$ _____ SE_{Est} _____ Interval:_____

9. What if: A shopper randomly selects four bags of oranges each bag labelled 10 pounds. The bags weighed 10.2, 10.5, 10.3, and 10.3 pounds. Assume the distribution of weights is Normal. Find a 95% confidence interval for the mean weight of all bags of oranges. Interpret your result.

10. In finding a confidence interval for a random sample of 30 students' GPAs, one interval was $(2.60, 3.20)$ and the other was $(2.65, 3.15)$

    a. One of them is a 95% interval and one is a 90% interval. Which is which and how do you know?

    b. If we used a large sample size $(n = 120$ instead of $n = 30)$ would the 95% interval be wider or narrower than the one reported here?

11. State whether each of the following changes would make a confidence interval wider or narrower. (Assume nothing else changes.)
    a. Changing from a 90% confidence level to a 99% confidence level.
    b. Changing from a sample size of 30 to a sample size of 200.
    c. Changing from a standard deviation of 20 pounds to a standard deviation of 25 pounds.

Table Number:_____ Group Name: _____

Group Members:_____ _____ _____

Tests of Significance: Means
NBA SCORING

Prior to the 1999-2000 season in the NBA, the league made several rule changes designed to increase scoring. The average number of points scored per game in the previous season had been 183.2. Let μ denote the mean number of points per game in the 1999-2000 NBA season.

a. If the rule change had no effect on scoring, what value would μ have? Is this the null or alternative hypothesis?

b. If the rule change had the desired effect on scoring, what would be true about the value of μ? Is this a null or alternate hypothesis?

c. Based on your answers to part a and b clearly sate **H₀** and **Hₐ** using the proper notation.

d. The following sample data are the number of points scored in 25 randomly selected NBA games played during December 10-12, 1999.

196	198	205	184	224	198	243	235	200
206	190	140	204	200	197	191	194	196
190	195	180	200	180	188	163		

Use technology to calculate the mean and standard deviation of the sample data. **Use appropriate symbols** to denote these values.

Mean:

Standard Deviation:

e. Comment of whether or not the conditions have been satisfied for the validity of the one-sample t-test?

f. Using the statistics found in part d, compute the value of the *t*-test statistic from the sample.

$$t = \frac{\bar{x} - \mu}{\left(\frac{s}{\sqrt{n}}\right)}$$

g. Use technology report the **p-value** for this study. Use an α level of .05.

h. Interpret the **p-value** in the context of these data and hypotheses. Write a sentence or two summarizing your conclusion about whether the sample data provides evidence that the mean points per game in the 1999-2000 season is higher than in previous seasons. Include an explanation of the reasoning process by which your conclusion follows from the test result.

Table Number:_____   Group Name: _____

Group Members:_____   _____   _____

Tests of Significance: Means
CHILDREN'S TV WATCHING

In 1999, Stanford University researchers conducted a study on children's television viewing. At the beginning of the study, parents of third- and fourth-grade students at two public elementary schools in San Jose were asked to report how many hours of television the child watched in a typical week. The 198 responses had a mean of 15.41 hours and a standard deviation of 14.16 hours.

Conduct a test of whether or not these sample data provide evidence at the 0.05 significance level for concluding that third- and fourth-grade children watch an average of more than two hours of television **per day**. (This would be 14 hours per week.) Include all components of a hypothesis test and explain what each component reveals.

Who are the subjects of the study? _____

What is the sample?

What is the population?

What is the variable we are measuring?

What is the parameter we are concerned about, a proportion or a mean?

State your Hypotheses here:

H_0:

H_a:

Check the conditions for a *t*-test here:

Find the test statistic, *t*, and write it here:_____ (Be sure to show work below)

Find the p-value using technology and write it here: _____

Reject or not reject H_0 ? Explain.

Interpret your conclusion:

If you had used a .10 significance level, would your conclusion be different? Explain.

Adapted from *Activity 20-4: Children's Television Viewing* in Rossman, Chance. Workshop Statistics, Discovery with Data. John Wiley Publishers, 2012. Pp. 428-429.

A Deck of Playing Cards looks like this:

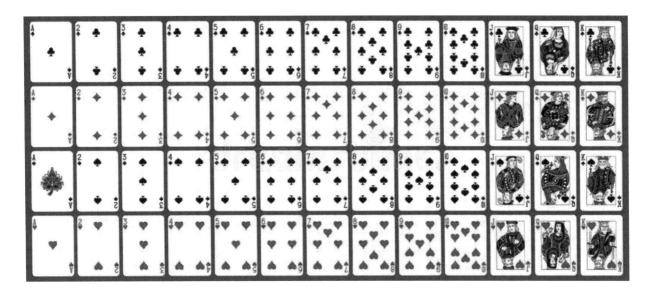

Notice the deck has four suits – clubs, hearts, spades, and diamonds. Each suit has 13 cards: ace, two, three, four five, six, seven, eight, nine, ten, jack, queen, king.